Dr. Eleanor's
BOOK OF
Common Ants

DR. ELEANOR'S

BOOK OF

Common Ants

✳

ELEANOR SPICER RICE,

ALEX WILD, *and* ROB DUNN

THE UNIVERSITY OF CHICAGO PRESS
Chicago and London

The University of Chicago Press, Chicago 60637
The University of Chicago Press, Ltd., London
© 2017 by Eleanor Spicer Rice, Alexander Wild and Robert Dunn

Published 2017

Printed in Canada

26 25 24 23 22 21 20 19 18 17   1 2 3 4 5

ISBN-13: 978-0-226-44581-6 (paper)
ISBN-13: 978-0-226-44595-3 (e-book)
DOI: 10.7208/chicago/9780226445953. 001.0001

LCCN: 2017026801

♾ This paper meets the requirements of ANSI/NISO
Z39.48–1992 (Permanence of Paper).

CONTENTS

PREFACE

I grew up imagining I might someday go to a faraway tropical forest as an explorer. I imagined that in faraway places big new discoveries were still possible, discoveries of great and hidden empires. I was lucky enough to go to such places and even to make a discovery here and there. But as I have gotten older, I have discovered something even more fantastic. I have discovered that great and poorly explored empires can be found not just in the deepest jungles but also in backyards. New species and even whole societies remain to be studied in the dirt beneath our feet.

Among the least explored empires are those of the ants. Ants live nearly everywhere. They do not appear to have made it to outer space, but it seems only a matter of time. Some kinds of ants have been very well studied, but just as for beetles, mites, spiders, and other arthropods, most have not. That is why the tropical explorer Andrea Lucky and I, along with a large number of colleagues, started a project called the School of Ants.

In creating the School of Ants, we aimed to give kids and adults around the United States (and now in parts of Italy and Australia) the wherewithal to go into their backyards and collect ants in order to document where ants of different species live. The project is new, but already the discoveries have been big. One boy in Washington State discovered an ant species living in his backyard that was thought to live only in the southeastern United States, for example.

But knowing which species live where is just a starting point. Having found the empires of the ants, the real challenge is to spend the time necessary to learn their ways. The good news is that for each of the most common ant empires in backyards, major discoveries are still possible. This is what I wish I had known as a kid. I wish I had known that instead of (or simply before) heading away to tropical forests to make new discoveries, I could have made them in my own backyard. But there is a catch.

The catch is that in order to make discoveries, one needs to know what is already known, where the last path ended and where a new one might begin. Before now, there has been no book describing what we know about the common ants of North America. Most of what is written about the common ants describes how to kill them (which is a shame given that most common ants do no harm and offer a great deal of benefit to our yards and even our homes). But here, Dr. Eleanor, an ant biologist from Goldsboro, North Carolina, tells their stories. These stories are fun, but they are also something more; they are a clear indication of where the paths end in our understanding of these common species. Some of the ants Dr. Eleanor writes about are relatively well known (see, for example, the fire ant), but most are not, and even those that are well known await major discoveries.

I wish I had had *Dr. Eleanor's Guide to the Common Ants* when I was young. I would have taken it—along with a bunch of glass jars, a shovel, a snake stick, and my other explorer's gear—out into the forest behind my house. With book in hand, I would have tried to add new information to the chapters. This is what I hope you do, because the truth is that each of Dr. Eleanor's funny stories about the most ordinary of our ants is just the beginning, and Dr. Eleanor needs your help to add new information. Perhaps you'll even help start a new chapter about a previously undiscovered species! And so go forth, young reader, and see what you can find.

Rob Dunn

INTRODUCTION: WHAT'S THE BIG DEAL ABOUT ANTS?

Before you dive into the stories of individual ant species, let's start with some basic ant biology and a little natural history.

Ants surround us, occupying nearly every type of habitable nook and cranny across the globe. Right now, ants snuggle up to your house, lay out their doormats in front of the trees in your yard, and snooze under park benches. Some even nest inside the acorns littering the ground!

We might not always notice them, but they're there, and they shape, literally shape, our world. Look at the colossal trees in our

forests. Ants like winnow ants plant the forest understory, ultimately contouring the distribution of plants in the undergrowth and even of those giant trees. Other ants help turn soil (more than earthworms in some places!), break up decomposing wood and animals, and keep the tree canopy healthy by providing rich soil nutrients for saplings, protecting tender leaves from predators, and regulating tree pest populations.

Ants creep across our yards taking care of business for us in much the same way. They eat termites and chase caterpillars out of our gardens. Even though some people think of ants as the tiny creatures that ruin their picnics, of the nearly 1,000 ant species living in North America, fewer than 30 are true pests, and fewer still actually can hurt us. Most ants spend their time pulling threads, stitching together the quilt of the natural world. Without these threads, the quilt would fall apart, becoming disconnected pieces of fabric.

In this book, you will meet our most common ants. Odds are you can see these ladies tiptoeing all around you. See how beautiful they are, with their spines and ridges, their colors and proud legs, each feature lending itself to the individual's task. See their work, how they help build the world around us as they move about our lives.

What's in an Ant?

Like all insects, adult ants have three body segments: the head, the thorax, and the abdomen.

Heads Up

Windows to the world, ant heads are packed with everything ants need to interact with their environments. With tiny eyes for detecting light, color, and shadow; brains for memory and decisions; mouths for tasting; antennae for touching and smelling, ant heads are one-stop shops for sensory overload.

petiole
thorax
antennae
gaster

Thorax

Ant thoraxes are mainly for moving. Although an ant's nerve cord, esophagus, and main artery course through its thorax, connecting head to bottom, thoraxes are mostly legs and muscle. Every one of an ant's six legs sticks out of her thorax, and when queens and males have wings, those wings stick out of the top of the thorax.

Abdomen and Petiole: One Lump or Two?

The abdomen is where all the action happens. This ant segment holds all the critical organs. Almost all of an ant's digestive system is packed into its booty, as well as tons of chemical-emitting glands, stingers and trail markers, the entire reproductive system, and most of its stored fat. Many ant species also have a crop, which is a special stomach in their abdomens that does not digest food. Ants use this stomach as a backpack to carry food back to the nest. There they share the food with their sisters by vomiting it back up and then spitting it into their sisters' mouths.

The first part of an ant abdomen is called the petiole. The petiole is that really skinny section between the thorax and an ant's big fat bottom, which we often call a gaster. The petiole gives ants their skinny waists and flexibility when they move around. A lot of people interested in identifying ants check the petiole first to see if it has one bump or two as a first step in determining the species.

Where's the Nose?

Unlike us, ants don't have noses. Instead, they smell and breathe with different body parts. To smell, they mostly use their antennae. To breathe, ants have little holes all along their body called spiracles, which they can open and close. When the spiracles open, air rushes into beautiful silvery tubes that lace the ants' insides and bathe their organs with the oxygen they need to survive.

The Ant Life Cycle

Like butterflies, beetles, and flies, ants grow up in four stages: a tiny egg, a worm-like larva, a pupa, and the adult stage that most of us recognize as ants.

After hatching, larvae molt several times.

Eggs

For most species, only the queen lays eggs that become workers. Most of the time, eggs are creamy-colored orbs smaller than the period at the end of this sentence. When queens fertilize the eggs, females hatch. When they don't, males hatch. Sometimes, like when the colony is just getting started, queens lay eggs called trophic eggs. Delicious and nutritious, workers and developing larvae eat trophic eggs and use that energy to help the colony grow. Occasionally, workers lay eggs, but queens and other workers sniff about the nest and gobble those eggs as soon as they find them, no bacon needed.

Larvae

The only time an ant grows is when it's in the larval stage. Most ant species' larvae look like wrinkly grains of rice or chubby little maggots. Fat pearlescent white tubes with plump folds and wormy mouths, ant larvae are the chicken nuggets of the insect world. When colonies get assaulted by other

Workers tend eggs and larvae.

ants or insects, these little chub monsters are usually the first to go. Because they have no legs, they can't run away, and they make easy meals for anyone able to break into the nest.

Unable to feed themselves, larvae sit like baby birds with their little mouths open, begging for workers to spit food down their gullets. When workers give larvae special food at just the right time, their body chemistry changes and they grow up to be queens. As larvae grow, their skin gets tight on their bodies. They wiggle out of the tight skin as if shedding an old pair of blue jeans, revealing a brand new, bigger skin underneath. Never wasting the chance for a snack, workers squeeze whatever liquid remains from those discarded larvae skins and sometimes feed the solids back to larvae.

If you look at older larvae under a microscope, you'll see sparse hairs jutting out of their supple flesh. I know a scientist who wanted to find out why they have these luxurious locks, so he gave larvae haircuts and watched what happened. The verdict? Shorn larvae fall over like little drunk sailors. They need their hair to anchor them to surfaces.

Some ant species, like this one, spin silken cocoons.

Pupae

When larvae grow big enough, they quit eating and get really still. The larvae of some ant species will spin a silken cocoon around themselves to have a little privacy during this pupa stage; others will just let it all hang out. Although they don't look like they're doing much during this stage, their bodies are changing and shifting around inside that last larval skin. They're developing legs and body segments, antennae and new mouthparts. They're turning into the ants we recognize.

When they're ready, they squeeze out of their last skins, emerging as full-grown ants. At first, they tumble about like baby deer, unsure on their legs, soft and pale like the larvae they used to be. But after a while, their skins darken and harden, their step becomes surer, and they begin their work as adults.

Workers

The two most important things to remember about ant workers are the following: First, all workers are adult ants. Once an ant completes its metamorphosis (when it is recognizable as an ant), it will never grow again. When you see a little ant, it's not a baby ant; it's just a species of ant that is really small. Second, all workers are female. Workers do nearly every job, so pretty much every ant you see walking around is a girl. While queens get the colony rolling and keep it strong by laying eggs, workers get the groceries, keep intruders out, take out the trash, feed the babies, repair the house, and more. When we talk about ant behavior and the special characteristics of ants in this book, we're talking about the behaviors workers exhibit in the natural world, since they are the colony's only contact with the outside world.

Queens

Despite their regal moniker, ant queens are mostly just egg-laying machines. When queens first emerge, they usually have wings, but after they find a lucky someone (or someones) and mate, they rub off their wings, let their booties expand with developing eggs, and go to town eating food and popping out eggs. Protected deep within the nest, workers feed queens and keep life peachy for them so they can produce healthy eggs for the colony.

An ant queen, with her large gaster, is ready for egg laying.

Males

Male ants are easy to discount because they don't seem to do too much around the colony. Unlike their more industrious sisters, male ants refrain from cleaning up around the house, taking care of the babies, going out to get food, or keeping bad guys out. The one thing male ants do for the colony is mate with queens.

To date, scientists have spent very little time studying male ants. But these mysterious and weird-looking creatures invite a closer look. Compared to their sisters, most male ants have tiny heads and huge eyes. Often, they look like wasps. Nobody knows for sure what the boys do when they leave the nest. What are they eating? Where do they sleep? Why doesn't anybody seem to care? As you read this, researchers are trying to finally put an end to our male ant ignorance.

What's in an Ant Colony?

Many different types of ants will nest in pretty much any type of shelter. While fire ants push up their great earthen mounds for all to see, acrobat ants might have their mail delivered to a tiny piece of bark on a tree limb, and winter ants scurry down inconspicuous holes in the ground to their underworld mansions.

While ant nests differ greatly, when you crack one open, you'll most likely find lots of workers (the ants we most often see in the "real world"), a queen (many species have several queens), and a white pile of eggs and larvae.

Most ants carry out the trash and their dead, piling them in their own ant dumps/graveyards called middens. A good detective can learn many things from going through someone's trash. If you examine a midden, you can get a good idea of what the ants have been eating and whether or not the ants are sick or at war with other ants. You'll probably discover bits of seeds and insect head capsules stuffed in with the dead ants. When tremendous numbers of dead ants litter the piles, it's likely the colony is sick or warring with other ants.

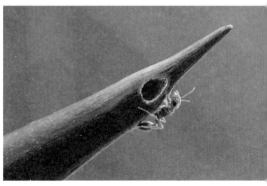

Back inside the nest, the ants busy themselves with their daily anty lives. You can take some cookie crumbs and call them out to you. See how they sniff the earth with their antennae, each one a living being experiencing the world, doing its special job. Watch them communicate, following one another under blades of grass and around pebbles, stopping every now and again to touch one another's faces, clean their legs, investigate their surroundings.

Ants nest in the ground, in thorns, and anywhere else.

Ants saturate our environment, from our homes to the sidewalks, city streets, and forests spread all around us. They are our neighbors, our friendly fellow citizens working away as we work. It's time we introduce ourselves.

01 CARPENTER ANT

SPECIES NAME: *Camponotus pennsylvanicus*

AKA: black carpenter ant

SIZE: 7–13 mm (0.25–0.51 in.)

WHERE IT LIVES: Black carpenter ants prefer to nest in living, standing trees but will also nest in logs and wood in human structures. They are widespread in the Northeast and Mid-Atlantic regions and are found throughout the Southeast (with the exception of Florida) and into the eastern Midwest region.

WHAT IT EATS: Omnivores, black carpenter ants eat protein foods, including other insects, as well as sugary foods, like the honeydew produced by aphids.

The black carpenter ant is one of the United States' largest and friendliest ants. Because of their size and pleasant disposition, they make excellent ambassadors between the ant and human worlds.

You can identify black carpenter ants by their size (BIG) and the light dusting of golden hairs on their head and thorax that settles on their abdomens. Unlike workers in some other ant species, black carpenter ant workers vary in size and shape within the colony. Between a quarter-inch and a little over a half-inch long, a small black carpenter ant can comfortably stretch over a plain M&M, and a large one can just about reach across a dime. Colonies have between about 350 to almost 2,000 workers, which, depending on worker size, works out to be almost 200 dollars' worth of dimes banging around inside those trees or, more deliciously, up to 40 bags of M&M's.

Breakfast for Ants

When I was little, I would take my breakfast crumbs out to my front yard to feed the black carpenter ants living in the willow oak trees. I would build little piles of bacon and toast for them on top of oak leaves and wait for them to lumber out from holes hidden in the bark at the bases of the trees.

I loved those ants. I was fascinated by the way they walked around like miniature black horses, exploring their way with their elbowed antennae, stopping every now and then to gently tap their sisters and give each other waxy kisses. If I pressed my ear against the tree near their entryway, I could hear them crackling about their business inside. If I sat still, they would come up to my hands and gingerly pick crumbs off my fingers. If I picked one up, she would explore my arm and shirt. If I squeezed her, she would give me a pinch with her tiny jaws. It never hurt.

They're called carpenter ants because they are particularly good at woodworking. They like to nest in living, standing trees using their sturdy mandibles to excavate tunnels and rooms in dead limbs or in dead wood in the tree's center. Many people see black carpenter ants living in their trees and think the ants are killing the trees. But black carpenter ants actually have a history of helping trees. They have an appetite for tree pests like red oak borers, and they spend a lot of their time foraging around their home, plucking pests off the bark. The trees housing my carpenter ants 25 years ago are still standing today.

Because of these woodworking skills, some people see carpenter ants as household nuisances. While black carpenter ants can make

their tunnels in the wood of people's homes, they often point homeowners to bigger problems: damp and rotting wood from a leak or drip or other pests living in that wood. When wood becomes soaked through, carpenter ants can easily use their jaws to snap it away and bore their tunnels. If homeowners keep their wood dry, carpenter ants will usually stick to the trees. That is, unless the homeowners have pests like termites or wood beetles snacking away inside their walls.

Sometimes black carpenter ants will happen upon such a treasure trove of food and set up camp right next to their grocery store. Can you blame them? Haven't you ever dreamed of living next to your favorite doughnut shop or fried chicken restaurant? Instead of attacking carpenter ants for living in your walls, use them as helpful guides to identify the real problem.

Ant Speak: Decoded

I used to think my carpenter ants might like some of my bologna sandwiches from lunch, but I couldn't get as many takers at lunchtime as I got early in the morning. That's because carpenter ants are mostly night owls, foraging from dusk until dawn. Black carpenter ants have pretty good vision for ants, and they use that vision to help them take shortcuts from their house to food in the early morning and when the moon is out.

When they aren't following their sisters' chemical trails, they remember landmarks like pebbles and sticks to help them find their way home. These landmarks save time for black carpenter ants, who can sometimes forage up to 100 yards from their nest. That's the human equivalent of walking more than 11 miles for food. On new moon nights, when it is totally dark, black carpenter ants take no shortcuts and feel their way through the night, keeping their bodies close to structures.

When carpenter ants find food, they run back to the nest, laying a chemical trail behind them. Once inside the nest, they do an "I found something awesome" dance to get their sisters awake and excited enough to follow them. The hungrier the ants, the more vigorous the dance. The excited sisters then rush out of the nest in search of the chemical trail that will lead them to the food. Carpenter ants, like many other ant species, have little built-in knapsacks called crops inside their bodies. They fill these crops with liquid food to take back home. When they meet their sisters on the trail, they stop and have a little conversation that goes something like this:

ANT HEADING OUT TO FOOD: "Hey, what's up?"

ANT RETURNING FROM FOOD: "Are we from the same nest?" (They check

this by tapping each other on the head with their antennae to see if they smell alike.)

HEADED-OUT ANT: "Yeah, but I'm not sure what I'm even doing here. I'm just following this trail." (She moves her tapping antennae closer to her sister's mouth.)

RETURNING ANT: "Oh, wow! I should have told you earlier. Some kid spilled his Dr. Pepper down the street, and it is DELICIOUS. Everybody's over there now drinking it up. Want to try?"

HEADED-OUT ANT: "That sounds awesome. Of course."

Returning ant spits a little droplet from her crop into headed-out ant's mouth. Headed-out ant drinks it and agrees it is awesome. Awesome enough, in fact, to continue running down the trail.

When I was a child, I saw black carpenter ants having these sorts of conversations all the time and thought they were kissing. When I grew up, I learned that I already knew much about black carpenter ants from watching them as a child. Their colony size, where they nest, and how they eat have all been scientifically dissected and explored as thoroughly as the ants themselves explore the dark tunnels of their homes. Scientific papers explain how they talk to each other, when they're awake, and why they don't want bologna on hot summer afternoons. Every delicate golden hair on the black carpenter ant's rump has been counted and cataloged. These discoveries took many decades to document. All of them can be remade any morning by each one of us, holding our breakfast crumbs, waiting patiently in our front yards.

02 PAVEMENT ANT

SPECIES NAME: *Tetramorium* sp.E (the ant formerly known as *T. caespitum*)

SIZE: 2.5–4 mm (0.1–0.2 in.)

WHERE IT LIVES: Pavement ants most often nest under bricks or pavement, but they are also found in grassy areas near sidewalks and even in extreme environments, like salt marshes. They are heavily concentrated in Washington and Ohio, sprinkled throughout the Mid-Atlantic region, and occasionally found in the Southeast region.

WHAT IT EATS: Ultimate opportunists, pavement ants eat anything from dead insects to honeydew, a sugary food produced by sap-feeding insects like leaf hoppers. They also dine on pollen and food in your kitchen and garbage.

Wars happen across America every spring. Just as the trees begin to give us that first peek of color and the sun warms us enough to stretch our legs and venture outdoors for a look around, the animals begin stretching their legs, too.

Pavement ant battles are brutal. Here, four workers tear apart an ant from another nest.

Each spring, ants poke their antennae out of earthen holes, getting a feel for their new year on the beat. Workers of the pavement ant species (*Tetramorium* sp.E— although the pavement ant is common, scientists have yet to give this species a real name) push out of their nests with a mission: to establish their neighborhoods before ants from other nests nudge in and squeeze them out. These ladies are territorial, and they don't like any other ants walking on their turf. When they first emerge in spring, all the previous year's boundary lines have been wiped away with winter, and all bets are off. They redraw their property lines with warfare so gruesome it would make Attila the Hun blush.

Pavement ants are built for battle. At three-sixteenths of an inch, workers are about half as long as one of your shirt buttons is wide. They are dark reddish-black and have antennae that bulge out at the tips, making them look like they're waving little clubs from their foreheads. They have tough, armor-like skins called exoskeletons that can withstand the knocks of war. If a pavement ant were the size of a dog and you could get a good close-up look, you would see a beautiful landscape. Their faces and bodies are covered with hilly peaks, rivers of grooves and hairs, and they have two little mountains of spines poking out from their backs toward their rear ends.

Where neighborhoods overlap, huge numbers of workers from each side collide. They furiously drum one another's heads with

their antennae; they rip one another apart with their mandibles. They'll separate an individual from the pack and close in around her, gnashing at her body with their jaws, grabbing her with their claws, turning her into ant dust. These ants mean business when it comes to setting boundaries. After the melee, the carnage is astounding. Thousands of ants litter sidewalks across the country, a jumbled dark line of body parts and pieces that blow around in the wind.

When they aren't out cruisin' for a bruisin', pavement ants move along slowly compared to other ant species, as though they don't have anything to do in this big old world but go for a walk in nature. They won't sting you, and they aren't easily spooked. Whereas some ants shoo away quickly, pavement ants usually continue to bumble along unbothered.

Pavement ants are not native to the United States, but they are one of the most common species around. They sailed over here in ships from Europe more than 100 years ago and flourish in the stone-slab environments of modern cities. They most often build their nests under bricks and in sidewalk crevices and will eat everything from sugary foods to dead insects to flower pollen to human garbage.

Sometimes, pavement ants act like miniature farmers. They collect seeds from plants and accidentally plant them by burying them in their nests. They also tend insects called plant hoppers like dairy farmers tend cows, "milking" them for honeydew, a sugary food the plant hoppers produce. If a plant hopper predator comes lurking around, pavement ants will pick the plant hoppers up in their mouths and carry them down to their nests, where they'll wait out the trouble. Pavement ants also keep interlopers off their property and will wipe out any ant nests that try to pop up on the homestead. But this is all during peace time.

Back to spring. The birds are practicing their songs, and you and I are hopping off the school bus, picking up lucky pennies, walking our dogs, or going to get coffee on our sidewalks that zig and zag from New York City down to Florida, across Tennessee, the Dakotas, and Wyoming, all the way to California. Each day, as we walk around in our world, the human world of sidewalks that point us to and from where we want to go, we are also walking over the world of the pavement ant, with devastating wars, property disputes, and peace times filled with farming and baby making. Their world is so similar to ours, so close to us, that we step over it every day without noticing how unusual these ants are.

03 ODOROUS HOUSE ANT

SPECIES NAME: *Tapinoma sessile*

SIZE: 2.25–3.2 mm (0.09–0.13 in.)

WHERE IT LIVES: Odorous house ants nest indoors (under sinks and doormats and in insulation and dishwashers) and outdoors (under rocks and in garbage cans, potted plants and exposed soil). They are heavily concentrated along the West Coast and in the Southwest region; found in the eastern Midwest states; and sprinkled along Florida and the Northeast and Mid-Atlantic regions.

WHAT IT EATS: Odorous house ants eat honeydew, a sugary liquid made by small, sap-feeding insects like aphids and scales, and other sugary food left out by humans. They also eat dead insects and spiders.

People across the United States call me all the time to tell me they have ants in their houses. It's one of my favorite parts of knowing a little bit about insects. From my grandmother Ina down in Opelika, Alabama, to my good friend Ariana out in Los Angeles to my friend Sarah's grandmother's friend up in Baltimore, the call is always the same: "Help me! I'm under attack! I've got ants in my kitchen!"

I love these calls because they make me feel like a real live wizard. Here's why: Across the United States, there are only three or four types of ants that often wander into people's kitchens. By asking a few questions, I can usually narrow the identity of the particular trespassers down to the species through a process of elimination. It's simple, but it seems like magic to the people who are calling. To let you in on the secrets of my sorcery, here is the phone conversation I had with Sarah's grand-mother's friend (SGF):

SGF: "Help me! I'm under attack! I've got ants in my kitchen!"
ME: "Are they big or little?"
SGF: "They're tiny!"

Clue 1: They are tiny. Now I know she doesn't have big carpenter ants or the less probable field ants. She also doesn't have Asian needle ants.

ME: "What color are they?"
SGF: "I gotta look at them? Hold on. I gotta get my reading glasses. Hold . . . on . . . OK! They're black!"

Clue 2: They are black. So, Sarah's grandmother's friend doesn't have pharaoh ants or fire ants. Plus, she probably doesn't have the brown Argentine ants. One more answer and I'll know what she has in her kitchen. Time for my big finish.

ME: "Here's what I want you to do. I want you to squish one. I want you to roll it between your fingers and put it up to your nose and sniff it."
SGF: "I'm sorry, what?"
ME: "Just do it. Tell me what it smells like."

Sarah's grandmother's friend squishes. She makes the I'm-squishing-an-ant sound people make, which comes out as a mix between "ooh!"

(fun!) and "eew" (gross). The result of this squish-and-sniff will tell me whether she has little black ants (about half the size of a sesame seed) or odorous house ants (a little bigger than a sesame seed).

SGF: "It smells . . . it smells good! It smells!"

Clincher: They have an odor. Like most people with ants, Sarah's grandmother's friend has odorous house ants partying in her kitchen. Their telltale smell gives them away. She's a lucky lady. Neither dirty nor dangerous, this top home pest—also known as the sugar ant—can provide hours of entertainment for anyone willing to share space with them. Follow them home to see how they bunk! Put out food and see how long it takes them to find it! Lay an *E.T.*-style trail of snacks to shift their ant highways! Possibilities for fun abound.

Country Ant, City Ant

Unlike some of the ant species that pester people around the country (imported fire ants or Argentine ants, for example), odorous house ants did not migrate here. They are US natives. Named for a defensive odor they emit from their rumps that some describe as

"spoiled coconut suntan lotion," they nest in natural environments like the woods or in pretty much any manmade locale like potted plants, under doormats, or in cars. As with Aesop's country mouse and city mouse, "country" odorous house ants (those living in natural, wooded areas) and "city" odorous house ants (those living in manmade environments) lead different lifestyles.

In the country, odorous house ants play an important role keeping the earth a clean, green machine. They work in concert with other forest bugs to keep tree canopies healthy and ensure a proper ecological balance with plenty of species hanging around. They also help accelerate decomposition and promote nutrient flow by eating dead insects and animals and nesting in and under rotting wood, in acorns, and in abandoned insect homes.

Yes, out in the country, they live the quiet life and have small colonies of a few hundred to a couple thousand workers. But once they move into cities, odorous house ants go a little wild. Their populations explode, sometimes spanning entire city blocks, and they blanket lawns and kitchen counters with greedy scouts sniffing around for a sugar fix.

When we build cities, we also build the perfect environment for odorous house ants to go berserk. First, it's easy for them to find a job to help support their city lifestyle. Plenty of ant employers looking for work (aka scale insects and aphids) await in the trees we plant to line our neighborhood streets. These creatures depend on odorous house ants to protect them from ladybugs, tiny wasps, and lacewings, all aphid and scale predators. When odorous house ants show up, those predators split, enabling aphid and scale populations to soar. To pay for their security detail, aphids and scale insects provide odorous house ants with a sweet syrup called honeydew.

In the woods, odorous house ants compete with different species for places to set up camp. With acorn ants stuffing their homes into acorns, citrus ants pouring out from under tree bark, and acrobat ants peeking down from tree branches, odorous house ants make do

Odorous house ants milk aphids like cattle for a sweet honeydew reward.

wherever they can. But in the city, they can nest anywhere. Vacancies abound. From our garbage cans packed with odorous house ant–ready foods to the luxurious mulch we pile up around our homes to our kitchen floors, odorous house ants feast, raise babies, and have shindigs around us all the time. City odorous house ants can have many nests per colony with tiny superhighways of workers moving between them, distributing supplies from nest to nest. Some odorous house ant colonies can span a city block.

In the country, as conditions around their nests change, such as when a new, more dominant species comes to town or a big storm floods the area, odorous house ants move out. They generally move their nests every two weeks or so. This ability to pack up and move willy-nilly in the woods helps them cope with ever-shifting, human-made environments. Garbage day? Dumpster-living ants can saunter over to the grassy area. Dumping out those potted plants? Odorous house ants who had been living inside happily toddle over to the compost pile. Having many queens in the nest helps them split up without too many tearful goodbyes.

Roll Up the Welcome Mat

While I see odorous house ants in my kitchen as a happy surprise, I'm aware that not everybody (OK, probably not most people) shares my sentiment. It can be disconcerting to see eager sugarbears trundling across your Wheaties. After I conduct my wizardly identification, the response never seems to be: "What FUN!" It's almost always: "How can I get rid of them?"

Store shelves are packed with poisons designed to extinguish these ladies. However, knowing what we know now about odorous house ants, most of us can outsmart them. Be a detective. Stake them out. Follow them home to see how they are sneaking into your house. Then, eliminate the access point. We know that odorous house ants like to hang out in tree canopies and bushes, slurping up honeydew. Walk around your house and see if you have any bushes touching your walls or windows. Branches bridge the ants from their outdoor lifestyles to apartment living. Cut back those branches. Snoop out other ways they enter the house. For example, they sometimes sneak in through cracks and crevices. Seal those with caulk.

We know they love to nest in mulch. People often dump piles of mulch around their homes. Switch that out for rocks, which odorous house ants don't like as much. Or try aromatic cedar mulch, which smells gross to odorous house ants, at least for a little while.

Look where they're crawling around inside, too. We know odorous house ants like sugar and all the delicious little treasures abundant in human garbage, so don't leave food out and tightly seal garbage cans. But even if you try to get rid of these sweethearts, pay attention as you do. Because the truth is, most of what might be known about these ants hasn't been uncovered. Most of their tiny empire's treasures lay undiscovered. So, while I can tell you as much as I've told you about sugar ants, I can't tell you much more. When someone calls to tell me about their sugar ants, most of what they have to report is not just grievance, it is science.

And so, when Ina says, "They keep stopping and talking to each other with their antennae," or Ariana reports, "I left my Coke open and they found it in less than 30 minutes!" these are things I write down, things you might want to write down too.

04 WINTER ANT

SPECIES NAME: *Prenolepis imparis*

AKA: false honey ant

SIZE: 3–4 mm (0.12–0.2 in.)

WHERE IT LIVES: Winter ants nest deep in the soil near tree bases or in open ground, like lawns. They are heavily concentrated along the West Coast, sprinkled throughout Florida, and found in the Midwest, Northeast, and Mid-Atlantic regions.

WHAT IT EATS: While winter ants won't pass up an opportunity for a sugary snack, these ladies prefer protein-packed food, noshing on other insects not lucky enough to endure winter's chill.

Remember in *Alice in Wonderland* when Alice followed the white rabbit down its bunny hole? The hole was ordinary enough at first, but once Alice climbed in, she fell down and down until she came

to a completely different world. Holes like that rabbit's pepper the ground across the United States. If we were as small as ants, we could tumble repeatedly down into other worlds. Winter ants are the white rabbits of ants. Plunging down their holes gives us a peek into their truly extraordinary lives.

Unless you follow a winter ant home, its nest's entrance can be hard to find. About the size of a buttonhole, winter ant nests aren't a lot to look at on the outside. Inside, however, deep mazes of tunnels connect chambers all the way to the bottom. The nests can extend almost 12 feet deep into the soil. That would be the human equiva- lent of a class of second-graders digging a hole more than 1.14 miles down, deep enough that 150 school busses could be stacked end-on- end before reaching the surface.

All that depth serves a purpose. While most ants are active in the spring and summer, winter ants prefer the fall and winter. Soil tem- perature does not vary as wildly as the temperature above ground, so when winter's chill plummets to 33°F, the winter ant's nest is kept insulated by the earth and remains at a balmy 64–68°F. This heat is important because between 40–50°F, most insects develop a serious case of brain freeze, going into what bug people call a "chill coma," where their muscles stop working so they can't move. Underground, winter ants beat the ice. Above ground, they dig short "warming tun- nels" scattered around their nest. When they start to get cold walk- ing around outside, they run down into the tunnels and warm up.

Staying Out of Trouble

My mother always told me the best way to stay out of trouble is to avoid it. Winter ants are masters at avoiding trouble because they move about when trouble is fast asleep. From March to November, when most ant species scramble around gathering food and fighting one another for space, winter ants seal themselves tightly in their nests. When November rolls around and other ant species tuck themselves in for their winter naps, winter ants unseal their

nests and begin exploring the world. Because they are active when other ants sleep, they often miss the dangerous tides of invasive ants that can wipe out many other ant species. In this way, they can persist in areas inhabited by other inhospitable ants. If they do happen to meet an adversary, they spray a toxic chemical from their rumps that scares off or even kills the would-be contender.

How to Spot Them

At the beginning of winter, winter ants are hard to identify. Shiny reddish-brown with lighter yellow legs, they look like your everyday, run-of-the-mill ant. Early in the season, workers are about 0.1 inch long, just long enough to span the letter *t* on this page. But as the season progresses and winter ants stock up on food, they become easier to identify.

To say a winter ant has a big behind mid- and late season is an understatement. When workers eat their favorite protein foods like insects and a sugary substance produced by other insects called honeydew, they stockpile the calories in special fat cells in their bums. These fat cells can grow to be of tremendous size. Because they wad-

Sometimes called "false honeypot ants" for their large rear ends, winter ants use that extra junk in their trunks to survive summer underground.

dle around with swollen rumps at the end of the season, some people call winter ants "false honeypot ants."

To understand why they pack on the pounds, let's poke our heads down into one of their rabbit holes. Winter ant colonies survive underground all summer on their rotund sisters' fat. Their ample behinds are the world's best refrigerators. Because fat cells are part of living tissue, as long as the worker is alive, the fat won't rot, like dead insects stored in the nest would. And because the fat is already concentrated and high in calories, workers don't have to process it like they would other foods. Winter ants store enough fat in their portly posteriors to feed each other and all the babies that emerge as adults the following fall. When workers unseal their nests in the fall, they emerge as Skinny Minnies again.

Home Deep Home

Let's travel a little deeper down the rabbit hole. While its worker inhabitants live a couple of years at most, winter ant nests can exist for more than ten years. The older the nest, the deeper it is. If we were winter ants crawling down into our home, we would enter through a short hallway leading to the first room. Other than the pinhole of light shining through the entrance, the whole house would be completely dark. To get from room to room, we'd have to smell our way with our antenna. Our rooms would have domed ceilings, tall enough for a couple of us to stand on top of one another. Because we'd have clingy feet, we could even walk on the ceiling!

We might have a few hundred sisters—sometimes up to 10,000— living with us, so every now and again, we'd bump into one of our sisters and give her a friendly tap with our antennae. If she seemed hungry, we might spit up a bit of food for her to eat. If she seemed dirty, we'd help clean her with our mouths and antennae.

It might take us a long time to get all the way to the bottom of the nest. Remember, winter ant nests are at least the human equivalent of a mile deep. Our older sisters would live in the upstairs rooms, and our younger sisters would live with our mothers deep down. Our queen mothers would wander around the bottom of our nest in the dark laying their eggs. Our younger sisters would help feed the babies and keep them clean, while our older sisters would gather food for us.

Life Underground

If we were winter ants, we would not be able to hear well, and it's quiet so far underground anyway. We wouldn't hear children running over us or leaves falling on our entrance. We wouldn't know somebody's dad's car just parked next to our own driveway. Beneath the roots, we wouldn't get wet when the sprinkler showers over our home and across the lawn in the summertime. We wouldn't hear the thud of the family dog flopping right on top of us to gnaw on a tennis ball. But it would all be happening above us, all over the United

States. If we were winter ants, we'd miss out on a lot of the fascinating lives of people. We're lucky we're not winter ants. We're people, active all year long, and able to understand and delight in the winter ant's secret wonderland, deep below our feet.

05 ASIAN NEEDLE ANT

SPECIES NAME: *Brachyponera chinensis*

SIZE: 3.5 mm (0.14 in.)

WHERE IT LIVES: In forests, Asian needle ants nest in rotting logs, under leaves and mulch, and under rocks. In human environments, Asian needle ants can nest anywhere from potted plants to under doormats, in landscaping materials or under dog bowls. They are sprinkled mostly along the Southeast but have also been spotted in Wisconsin.

WHAT IT EATS: While they love termites, the Asian needle ants will eat pretty much anything they can find, from dead insects to other ants to human garbage.

The Asian needle ant (*Brachyponera chinensis*, formerly known as *Pachycondyla chinensis*) reminds me of a ninja superspy. Sleek, sneaky, and all dressed in black, ninjas (at least in bad movies) are masters of disguise and inevitably up to no good. The same holds true for the

furtive Asian needle ant. This stealth operative is sneaking across forests and backyards throughout the eastern United States.

Asian needle ants originally sneaked into the United States from Japan. Nobody knows how they got here, but they have been moving log to log since at least the 1930s. Slender, shiny, and black with lighter orange legs, Asian needle ants look like they are dressed for subterfuge. At about 0.2 inch long, one worker is almost as long as a kernel of unpopped popcorn.

Asian needle ants aren't fussy when it comes to where they make their homes. In the woods, Asian needle ants nest in logs or under rocks and leaves. Sometimes their nests look like caverns connected by tubes and stuffed with eggs and ants. Other times they look like nothing more than a group of ants hanging out. Around human structures, they nest anywhere from potted plants to piles of mulch, and even underneath doormats. Colonies can have anywhere from a few dozen workers to a few thousand, and those workers can live in one big nest or in many small ones.

The Asian needle ant's distinctive walk is a dead giveaway of its identity. While some ant species lift their legs high and prance around or stomp their way to and from their nests, Asian needle ants hunker down close to the ground and creep with deliberate, stealthy steps. Like ninjas, they move alone; they never follow the trails of their sisters—they don't know how.

It's easy to confuse Asian needle ants with wood ants, as both are medium to large in size and black, but one distinguishing characteristic separates Asian needle ants from wood ants (and ninjas, for

that matter): They are clumsy and terrible climbers. If you trap an Asian needle ant in a glass jar, she won't be able to climb to the top like other ant species and will instead wander helplessly around the bottom of the jar or run in place like a startled cartoon character. Be careful if you try that trick because Asian needle ants can sting the tar out of you.

A Stinging Sensation

I'm an entomologist who spends a lot of time studying Asian needle ants, but I'd heard of the horrors of their sting even before experiencing it for myself. The day the first sting happened, I was digging around with bare hands into a log I hoped was infested with Asian needle ants. It was. As I reached into the log to pull off a chunk of wood, I accidentally closed my hand on the nest. A startled worker stung my palm. Because I had read that two to four times as many people are allergic to Asian needle ant venom than are allergic to honey bee stings, my subsequent alarm seemed justified. Based on those reports, I was afraid my hand might fall off, but nothing like that happened.

At first, I felt a slight burning sensation right where she stung me. About an hour later, the burn spread out to an area about the size of a quarter around the sting, and it began to feel a little like being stabbed with pins. For the next two weeks, this flash of sharp pain followed by a dull nerve ache occurred every time I touched the area around the sting. For those of us not allergic to Asian needle ants, that's the worst part of their stings. I've been stung innumerable times since then, and it's almost always the same.

While Asian needle ants have pricked me many times over the last few years, I don't blame any of them for doing it. Unlike the war-mongering fire ants, which eagerly attack en masse, stinging anything they can get their angry little tee-hineys on, Asian needle ants prefer a more peaceable lifestyle and sting only in self-defense as a last resort. Every one of my stings occurred when I put pressure—whether on purpose or by accident—on a worker, so she poked me with her stinger to get away.

What's for Dinner?

Most of the time, Asian needle ants use their stingers to subdue their favorite food: termites. Watching an Asian needle ant around termites is like watching me at an all-you-can-eat buffet. She gets very excited, running around grabbing every termite she can. Practically defenseless, termites have thin, soft exoskeletons and are juicy treats for any meat-loving insect. When an Asian needle ant stings a termite, she grabs it in a bear hug and jabs her stinger deep inside. Her venom paralyzes the termite but does not kill it. By keeping the termite alive, she can stockpile it in her nest without worrying about it rotting before she and her nestmates get a chance to eat it.

Termites make easy meals for Asian needle ants.

Asian needle ants love termites, but they aren't picky eaters. If you see one out and about, she is probably scavenging the ground for other ants, dead or dying insects, or even human garbage. Unlike other ant species, Asian needle ants do not follow foraging trails. If a worker finds food too big to bring back to the nest, she will run home and tap one of her sisters imploringly on the head. Her sister will respond by folding up in the fetal position. The forager will then pick her sister up, tuck her under her body, and creep as fast as she can back to the food. Then they'll work together to bring the food back. If two aren't enough for the job, they'll carry more sisters over to help.

Like ninjas, Asian needle ants are masters of disguise and sometimes sneak into other ants' nests undetected, killing workers. They steal back to their own nests with ant bodies in their mandibles. Then the feast begins.

Their Covert Operation

The Asian needle ant ninja army has a stealthy mission worthy of our attention: to steadily disassemble ant communities in forests across the United States. When Asian needle ants move into a forest, other ant species like winnow ants, acrobat ants, little black ants, and thief ants all pack their bags and move out, pulling the forest apart at the seams. Asian needle ants make life miserable for other species. They eat them or their food (or both) and take up space native species use for nesting.

"Why should we care if a few ant species go missing?" you might ask. "All ants do is ruin my picnic! We could do with fewer of them anyway!" Let's think about this in another way.

Picture a car factory where every employee has his or her special job. One person puts on the windshield wipers, another, the wheels, while another is responsible for the engine and another adds the finishing touches like door handles. They all work together to build a beautiful, well-oiled machine. Suppose one day the company hires a new employee to add a new gadget in the car. This employee gets paid a lot of money, so much money the company has to lay off the windshield wiper person, the wheel person, and the engine person. They even kick out the door handle guy.

Here's the problem: While this new employee is really good at gadgets, he doesn't know anything about engines or door handles. You can forget the windshield wipers and wheels. What kind of car will this car factory produce without the other employees? One that won't even roll!

Our forest is a lot like that car factory. While some of us might think of ants as pesky, most ant species help keep the world rolling along. In fact, of the more than 30,000 ant species in the world, less than 0.3 percent are pests. The rest have valuable jobs, and we need them to show up for work each day. Take some of the species Asian needle ants displace, for example. Acrobat ants and thief ants help keep the forest canopy healthy by regulating tree pest populations and protecting tender leaves from leaf predators, as well as contributing to the nutrient-rich soil that young trees need to grow and survive. Winnow ants move seeds across the forest floor, controlling the distribution of forest plants and promoting healthy forest herb diversity and growth. Little black ants turn the soil, aerating it to keep the trees and shrubs happy. Taking away all these species and replacing them with just Asian needle ants can spell trouble for forest health.

Asian needle ants, those little ninjas infiltrating our turf with their clandestine movements and veiled operations, are a force to be reckoned with here in the United States. Our ants and forests aren't prepared to battle this stealthy foe. They need us to help them fight back. You and I can use our knowledge to spot them and work to kick them out. We can let people like the folks at the School of Ants know when we find them, so they can track their movements across the United States and research ways to keep them at bay. Asian needle ants might be ninjas, but you and I are part of a citizen army. Together, we can beat them.

06 WINNOW ANT

SPECIES NAME: *Aphaenogaster rudis*

SIZE: 3.81 mm (0.15 in.)

WHERE IT LIVES: Winnow ants prefer to nest in rotting wood but will nest anywhere from soil in open areas to human garbage. They are distributed primarily throughout the eastern half of the United States but have also been found in Wyoming and Colorado.

WHAT IT EATS: Winnow ants eat the tasty outer coatings of seeds, and other insects like termites. They also like sugary foods.

Aphaenogaster rudis sounds more like an unsavory medical condition than one of the coolest ant species in North America. The name doesn't roll off the tongue like "sugar ant," "carpenter ant," "pavement ant," or "fire ant." So, for the purposes of familiarizing you with

one of the best residents on your block, we'll give *Aphaenogaster rudis* a nickname: the winnow ant.

Winnow ants are among the most elegant-looking ants around the forest and in your backyard. With their long legs and slender reddish-brown bodies (leading to the nickname they share with their cousins: thread-waisted ants), they pick their paths delicately across the ground like rusty ballerinas. Each medium-to-large worker measuring at about 0.15 inch can just cover the date on a quarter. Although they prefer to nest in decomposing stumps and logs, winnow ants can make the best out of any situation, building their homes in open soil, beneath rocks, and even in human garbage.

Cracking open forest logs in the spring often reveals a hidden world of winnow ants.

With one queen and up to 2,000 workers, a winnow colony could easily pack a stadium for an ant rock concert.

Beyond their refined appearance and wide-ranging nesting habits, winnow ants have two qualities that set them apart from the rest of the ants: the helping hand they give forest plants and their ability to use tools.

First, let me tell you about their agricultural talents and the reason we call them winnow ants. Winnow ants have a special relationship with forest plants. We all know that many plants make seeds. Some plants produce seeds with a special coating called an elaiosome that's a lot like the hard candy coating on the outside of an M&M. The elaiosome has a special blend of flavors that is irresistible to winnow ants.

As they pick across the forest floor in search of food, winnow ants often stumble across these seeds. When winnow ants get a whiff of that elaiosome, they can't help themselves: They have to pick up the seed and carry it back to their nests. Once in the nest, winnow ants feed the outer coating of the seed to their young.

Winnow ants munch on a seed's tasty outer coating before planting the seed.

Unlike most of us, who prefer the chocolaty center of M&Ms, winnow ants eat only the elaiosome and leave the seed inside alone. When wheat farmers remove the chaff from wheat seeds, it's called winnowing. Likewise, winnow ants remove the chaff from forest seeds. Once the elaiosome is gone, the ants don't need the seed anymore, so they take it back out of their nest and deposit it on the forest floor or in fertile underground middens. There, the seed, no worse for wear, is free to sprout and grow into a happy forest herb. Having their elaiosome nibbled away for hungry ant babies does not hurt the seeds; in fact, it helps them. When ants take these seeds back to their nests, they in effect protect them from animals that eat the whole seed. Later, winnow ants "plant" the seeds (by discarding them outside the nest or in middens) far away from the seeds' parents. As a result, the newly planted seeds don't crowd their parents as they grow.

Seed planting is a successful business for winnow ants and the seeds they plant. Almost two-thirds of all herb seeds produced in the forest, such as from wild ginger and trillium, are picked up by winnow ants. Also, when winnow ants are removed from forests, some wildflower abundance drops by 50 percent. Seed planting also helps the ants. When winnow ants eat that candy coating elaiosome, they get all the nutrients they need to make more babies.

Farming isn't the winnow ant's only talent. Like other animals, from woodpecker finches to chimpanzees to humans, winnow ants use tools to gather food. When a winnow ant happens upon liquids too goopy to carry back to her nest, she goes out in the forest and collects bits of soil, leaves, and sticks. She takes these bits back to the newfound food and drops them right on top of it. These leaves and sticks become little plates for winnow ants. Workers bring the plates back to the nest for the colony to feast from like Sunday churchgoers at a pot-luck dinner.

Aphaenogaster rudis is a fancy name, but the winnow ant has earned it. How many of us have stood in the living city of a forest, awed by the architecture surrounding us? Somewhere, tucked into the hustle and bustle of creatures keeping the forest alive, creep winnow ants, rusty little architects helping shape everything we see on the ground floor.

07 BIG HEADED ANT

SPECIES NAMES: *Pheidole bicarinata, P. dentata,* and *P. tysoni*

SIZE: minors about 2.54 mm (0.1 in.), majors about 3.56 mm (0.14 in.)

WHERE IT LIVES: Big headed ants inhabit grassy, open areas and forests. Sometimes they move into people's homes. *Pheidole bicarinata* are heavily distributed in the Southwest, Southeast, and Mid-Atlantic regions and also sprinkled throughout the Northwest. *P. dentate* are distributed in the Southwest and Southeast regions and dotted throughout the lower Mid-Atlantic region. *P. tysoni* are sprinkled along the Southeast and Mid-Atlantic regions, up to New York's southern tip.

WHAT IT EATS: Big headed ants eat anything from sugary foods like aphid honeydew to dead animals, human garbage, and seeds.

The first time I met my husband Gregory's family, I was convinced he was adopted. He and his brother, Henry, look nothing alike. Where

Gregory's lean body stretches to more than six feet tall, Henry has the shorter, sturdier build of his parents. A thick mop of almost black hair flops over Gregory's brow; Henry's crown is lightly dusted with strawberry brown like their father's. Their teeth are different; their voices don't sound alike; even their fingers and toes are different. I tried to get his parents to confess Gregory's adoption, but they assured me he is their son. On holidays, I tried to trip up his aunt and cousins over the conditions of Gregory's birth, but everybody maintained that he is 100 percent, for sure, *not* adopted. Over time and after much superspy-like investigation on my part, I finally gave up and conceded that Gregory and Henry probably have the same parents. They just don't favor each other. Likewise, it took a similar amount of convincing for me to believe that majors and minors, the two types of workers in big headed ant nests, were not only related but sometimes almost genetically identical.

Big headed ants move about grassy knolls, sandy parks, and forested areas all across the United States, along the east and west coasts and all states in between. Three species out of many make the most common list: *Pheidole bicarinata*, *P. dentata*, and *P. tysoni*. They usually nest in dugouts of exposed soil or under rocks or other objects, but sometimes you'll find them in rotting wood or tree stumps or around the house. They have catholic appetites and eat sugary foods, dead insects and animals, human garbage, and seeds.

Big Heads, Little Brains

Whoever gave these ants their common name didn't get too creative. While the minor workers in big headed ant nests look like

regular, run-of-the-mill ants, the majors have tremendous heads. Like, huge. I mean, their onions can be bigger than their big bottoms, and sometimes they look like they'll topple forward at any moment. Unlike their smaller sisters, majors also have large snapping mandibles.

As with Gregory and his brother Henry, different physical characteristics offer different advantages. While Gregory's nimble fingers and wide hand span help him to play the piano, Henry's sturdier build makes him a great football player. With big headed ants, the minors' sprightly size helps them to squeedle in and out of the nest doing typical ant chores like feeding babies, taking out the trash, looking for food, and making sure the queen's all right. Their sleek bodies move easily between blades of grass and across fallen leaves and other obstacles they encounter as they navigate their world.

Majors, on the other hand, tend to be more linebackers than eggheads. Their big noggins packed with muscle, not brain, majors lend their extra brawn to cut up food too big for minors and to chomp down on intruders unlucky enough to stumble across their nest entrances. While the size differences between majors and minors can be astonishing, understanding how they got that way takes one's breath away.

Here's what's so cool about it: When queens lay a female egg, that egg has the potential to become either a major or a

Majors lend their brainier sisters some brawn.

minor. That's right. Majors and minors come from the exact same eggs and can have the exact same genetic makeup. Their physical differences emerge based on what their sisters feed them as babies.

So here's a baby big headed ant, a happy pearlescent porkie pie, mouth gleefully open, hoping somebody will come into the nursery and stuff it with more grub. For a certain amount of time in her life as a baby, if her sisters feed her lots of high-protein food (think smashed-up dead insects and Spam), a switch will flip in her grubby little body and chemicals will change her into a big headed major when she grows up. If her sisters give her just regular food (some protein but also sugary treats), no switch flips, and when she grows up she'll be a little-headed big headed ant minor. It's that simple, and that beautiful. And it all comes down to food. It makes me wonder about all those fried chicken dinners at Gregory's house.

Because workers do all the feeding, they can decide how many big heads they want knocking around the nest. For example, when other colonies move into the neighborhood, taking up resources like food and nest space from the big-heads, workers start pumping their babies with protein and making lots of soldiers to prepare the nest for battle.

Don't Quit Your Day Job

But the story of the big headed ants' jobs doesn't end with their size. Once they're grown, big headed ants, and ants in general, can't hold on to a job. Consider all the chores in an ant colony: baby feeding, grocery shopping, taking the trash out, keeping out the bad guys. Like people, who need to learn how to roll over, crawl, and then walk in sequence, big headed ant minors work different jobs as they grow up.

When they first emerge as adults, minors' antennae and mouth muscles are weak. I guess they didn't get much of a workout in their childhoods, sitting still snarfling down every scrap of food they could get their mouths around. Too weak to forage or feed their sisters, these youngest workers mostly wander around, getting their sea legs.

As they get older and learn the ropes a bit, minors start to pitch in, nursing their baby sisters with food their older sisters bring home and taking out the trash every now and then. In the sunset of their lives, they finally venture out of the nest, searching for food and defending the nest from intruders.

Something wonderful happens to big headed ant minors as they begin their foray into the outside world. Their brains change. Ants don't need to know a whole lot when they're wandering around in the darkness of the nest. Feed this, clean that. But once they step outside, whoa! Information overload! Outside the nest, ants need to *remember*. What does my nestmate smell like? How did I get here? How do I get home? Where do my enemies live?

It might seem like a pretty steep learning curve, but big headed ant minors have it all worked out. As they age, these ants' brains change to help them remember things. Chemicals in their bodies rewire their brains to help them forage better, and their brains (not their heads, but their brains), particularly the parts where they store their memories (called mushroom bodies), get much, much bigger. All the better to organize and store that information avalanche when they begin moving around the outside world.

Getting the Big Head

So Gregory's not really adopted. Final proof came for me at the birth of Greg's niece, Caroline, a miniature replica of my husband, down to her wide blue water-drop eyes and plump bottom lip. And big headed ant minors didn't adopt their broad-crowned sisters. They formed them, working within their splendid system, the elegant interplay of environment and chemicals and tissues, the resulting dance of living creatures in a living world, perfectly configured elements of nature. No fried chicken dinners needed.

08 SOUTHERN FIRE ANT

SPECIES NAME: *Solenopsis xyloni*

SIZE: 2.5–4.6 mm (0.1–0.18in.)

WHERE IT LIVES: Southern fire ants nest in mounds or flattened craters in open soil near moisture. They also can nest under carpets, in crawl spaces, or under rocks. Once common across the southern United States, they now live primarily in the Southwest.

WHAT IT EATS: Southern fire ants have a healthy appetite for pretty much anything, including dead insects, sweets, greasy foods, and sometimes seeds.

Here is a story about a villain, a stinging sensation, and a possible hero. It is the story of the Southern fire ant, known to scientists as *Solenopsis xyloni*. Easily confused in appearance and behavior with the red imported fire ant (*Solenopsis invicta*, that notoriously bad guy

Southern fire ants make quick work of a tasty grasshopper.

from the Southeast), Southern fire ants generally make their (much smaller) homes in small, loose dirt mounds in grassy openings or under rocks or boards in the southwestern United States.

Our story begins with a villain and a few simple facts about the birds and the bees. Actually, just one fact: birds and bees dread Southern fire ants.

As ground-dwelling predators, Southern fire ants can devastate ground-nesting birds like bobwhite quail. Although a quail adult outsizes a Southern fire ant worker 160,000-fold, Southern fire ant nests can have upward of 15,000 workers, who comb the ground with their pinchy mandibles and venomous stingers, swarming quail nests and devouring their much smaller chicks. In fact, Southern fire ants rank up there with coyotes, skunks, and badgers as one of the top four predators of bobwhite quail nestlings in Texas. But their bad behavior toward birds doesn't end with bobwhites. Southern fire ants also attack house sparrow chicks and endangered least tern babies. Killing Southern fire ant nests increases least tern populations because the ants aren't around to chomp away at their fuzzy little babies.

While the gruesome death of one of nature's most squeezably fluffy inventions (baby birds) may seem horrible enough, plants might argue they've got it even worse.

Let's be a plant for a minute, maybe a happy little bush. Here we are in our field, and life is pretty sweet. The sun is shining on our leaves, and our flowers are just beginning to open and say hello to the world. We smell so good, enticing fat little bees to come bury their fuzzy heads deep within our petals for a drink of nectar. Hello, bees! As they burrow in for a sip, they dust their heads with our pollen and transfer it to other flowers. We need these bees to visit us so we can do the one thing we need to do most in this big old world: reproduce. Make seeds.

Normally, we love ants. They keep the number of pests squiggling around on us stable so we don't get sick or depleted. But to us, these Southern fire ants are the *worst*. First, instead of controlling pests, they protect insects that can give us disease or suck us dry, harboring them against predators in exchange for a nectary treat the pests give them as a reward.

Lots of other ant species do the same thing, but not many are as good at it as Southern fire ants. Southern fire ants are such fierce protectors of their flocks of pestilence that they scare away anyone else who comes near us, including pollinators like bees and butterflies.

To make matters worse, these ladies rob nectar from our flowers. That is, instead of dusting their bodies with pollen and moving that pollen from flower to flower like bees do, these little jerks bypass our pollen when they drink up our nectar. That's right. Not only do they make us lousy with all those pests, they also prevent us from making seeds, our only task in life!

By now you might be thinking, these ants are just terrible. But I know one ant expert named Andrea who would disagree with you.

"I love Southern fire ants," she told me once. "They're so cute and shy. They always try to run away from you. Besides, they're native and they're always getting pushed around by those other fire ants."

This is the other side of Southern fire ants, the side that holds a valuable place in nature. They evolved with the natural environment across the southern United States, helping to regulate the balance of animals and plants in their natural home. Southern fire ants used to live all across the southern United States, but around the 1950s, red imported fire ants and their big, grumpy colonies began trouping across the Southeast, stomping out many Southern fire ant nests in their way.

Have you ever met somebody with a bad reputation and expected to dislike them, only to find out that they were actually pretty nice? As Andrea pointed out with her observation, this may be the case with the Southern fire ant; it's possible that we just haven't really *met* them yet.

With science, we only know what we have studied so far. We've studied Southern fire ants' bird-eating, plant-hurting behavior, so that's their reputation for now. But we have a lot left to learn about these insects. How has the environment changed since red imported fire ants came and Southern fire ants left? How do Southern fire ants behave around other ant species? How often are we mistaking the deeds of red imported fire ants for those of Southern fire ants? I'm sure you could come up with some good questions about Southern fire ants yourself.

We can build on and challenge what we know through exploration. You and I can form our own opinions based on information. Isn't that wonderful? With science, you and I can each meet these ants, ask questions about them, and discover for ourselves more chapters in the Southern fire ant's story.

09 LITTLE BLACK ANT

SPECIES NAME: *Monomorium minimum*

SIZE: 2 mm (0.08 in.)

WHERE IT LIVES: Little black ants can make their nests outdoors, in forests, or right in our backyards, often under rocks and tree bark. They are heavily distributed in the Southwest, Southeast, and eastern Midwest regions. They have also been spotted in the continental United States as far west as California, as far north as North Dakota, and as far northeast as New York's southern tip.

WHAT IT EATS: Little black ants eat a sugary liquid called honeydew, made by small insects called aphids and scales. They also eat dead insects, spiders, and your trash.

When I was little, we had a nest of what I later learned were little black ants under a cherry tree in our yard. In the thick of summer, tired of digging foxholes all over the yard, my brother Will and I

would follow them. We'd grip the trunk with our monkey toes and climb to the outside branches. We'd lean out as far as we could to see what they were up to. So long as we stayed out of their way, they never seemed to mind. They went about their business beneath the leaves and around the branches as we went about ours.

Little black ants are among our cutest "most common" ants. As their common and scientific names suggest, little black ants are much smaller than many of the other ants you see hanging around your house and yard. Their glossy sheen adds a touch of determination to their comings and goings. It's as if they take themselves too seriously, little polished wingtips toddling to and from their important ant business.

In our ant stalkings, Will and I quickly learned that, while they might be fun to watch, little black ants could be little jerk ants if we interrupted their work. Although small, these ants would jab their tiny stingers in our thighs and arms if we accidentally blocked their trails. It doesn't hurt much, but it's enough of a reminder to keep moving.

Little black ants bully other ants over food resources. Their colonies can number more than 2,000 workers, and when they get upset, they recruit their sisters in high numbers. When little black ant workers combine forces, their tiny stingers can pack a powerful punch to other ants. They put their best combat skills on display when they protect one of their favorite foods: the sweet nectar produced by sap-sucking insects such as aphids.

With mouths shaped like drinking straws, aphids live on plant leaves. They stab these straws into the leaves and suck out the juice like it's a big milkshake. They then turn that juice into honeydew that they excrete from their rear ends in droplets they hold high in the air, waiting for ants to come and get it.

To reach their favorite syrupy snack, little black ants travel in long lines up tree trunks and plant stems. They make the line by laying down scented pheromone trails. Even after their sisters are gone, the trail remains, a scented road to good food that they follow by waving their antennae back and forth over the path.

In addition to making delicious honeydew, aphids are tasty snacks for other insects like lacewings and ladybugs. But scrappy little black ants kick out other would-be diners from their honeydew buffets, even if those diners dwarf our tiny, shiny brawlers. Despite their Lilliputian size, little black ants' stingers and chompy mandibles can inflict more damage than other, larger ant species. With little black ants around to keep predators away, aphid numbers increase up to 10 times their normal abundance.

To little black ants, grains of sand are boulders.

Although up to the challenge in groups, when little black ants get caught alone, they have other options. Suppose a fire ant finds a little black ant hanging out on a root and decides to pick a fight. Instead of fighting back, our little black ant will "flag" her gaster (her abdomen), wagging it around in the air as if to say, "You'd better stay away from me! I mean business!" While she wags, she will release noxious toxins, hoping to repel the contender before she has to fight.

If booty-shaking fails, our little black ant will curl up and act dead, playing possum in the hope that the fire ant will think herself victorious and just go away. Sometimes, little black ants combine their individual possum-playing and group brawling behaviors to persist in areas with more dominant ant species. These little ladies can even push out fire ants trying to move into their neighborhood. They snack on fire ant babies as a reward for triumphant battle.

When Will and I watched the little black ants twine around our cherry tree on those hot summer days, fire ants had not yet made their march into North Carolina. Little black ants were the only game in town on that side of our house, with carpenter ants and field ants galloping through the front yard and high noon ants staking their claim to the hard-packed dirt and centipede grass in the backyard. Will's a grown-up lawyer now; his monkey toes spend the day in dressy shoes. The cherry tree was cut down 20 years ago, its last fat fruits still clung to the branches all piled up on the curb. But little black ants are still the same. When I find them greeting me on the walkways of campus or snaking across my porch, their shiny heads determinedly pushing forward, I fill up with the pleasure of seeing old friends. When we understand these elements of nature, get to know them by name and habit, we will always be surrounded by friends.

10 THIEF ANT

SPECIES NAME: *Solenopsis molesta*

SIZE: 1.5–1.8 mm (0.06–0.07 in.)

WHERE IT LIVES: Thief ants nest underground in forests and open, grassy areas. They also like to nest in human structures. They particularly like nesting near other ant species' nests. Thief ants are heavily distributed in the north-west tip of the United States, in California, in the Southwest region and sprinkled throughout the eastern half of the United States.

WHAT IT EATS: Tiny ants with big appetites, thief ants prefer protein, such as dead insects.

Back in the days of the Wild West, Jesse James and his outlaw gang were some pretty crafty dudes. They robbed everything from stage-coaches and trains to banks and homes. His bandit bunch crept into

towns and would hightail it out ahead of angry lawmen and WANTED posters bearing James clan faces. Imagine if the Jesse James family moved in right next door to your house! Many ant species across the United States face this predicament every day when thief ants come to town. Thief ants are the Jesse James gang of the ant world, and these bite-size burglars pickpocket and plunder anything they can get their little mandibles around, living lives of artifice that would make Mr. James sit up and take some notes.

Even though he was a robber and a murderer, Jesse James won the public's hearts, in part because he was easy on the eyes. Thief ants are no different. Whenever I stumble upon a thief ant nest or happen to lift a dead insect and find a bevy of thief ants, mid-snack, I always stifle a squeal. Thief ants are unbelievably, ridiculously cute.

Their size might play a big factor in their cuteness. At one-sixteenth of an inch, a thief ant worker could wander comfortably around in a lower case *o* on this page. Most often a golden yellow color, thief ant workers vary along the color spectrum all the way to amber. They have stingers, but they are too tiny to cause you any pain. They look like they wander around really slowly, but actually they're just super small. If you had a microscope, you could see that each antenna has a bulb on its end, and they bonk about as they feel their way to and from food. Much of that food, remember, is stolen, either from other ants or from you and me.

Thief ants get their name from their habit of setting up camp next to other ant species' nests. They love protein and stuff their bellies on dead insects, people food, and insect eggs. When the other ants bring home thief ants' favorite foods, those crafty little burglars sneak that food right on over to their own houses and feast. They've also been known to smuggle out other ants' babies, tasty snacks for greedy thief ants. When other species' colonies are weak or dying, thief ants aren't as sneaky. They run through the nests' halls like children running down the aisles of a Toys"R"Us on a shopping spree, eating their fill of dead and dying ants.

Their crimes and misdemeanors don't stop with the insect world: A thief ant will rob you blind if you don't watch out. Thief ants are opportunists, and they recognize that your kitchen is a wonderful opportunity for the biggest heist of their lives. Because they are so small, many people have a hard time figuring out how to keep them out of their pantries. The best way to keep thief ants out is to figure out how they're getting in. Once you do that, block their entranceway by plugging holes with some caulk or weather stripping and tell those thief ants there's a new sheriff in town.

Some people think Jesse James was like a modern-day Robin Hood and that many of his crimes were to benefit others. I don't know what Jesse did with all of his loot, but many of the thief ants' crimes against other insects surely do help us out a lot. For example, when they're not stealing from other ants, they love to eat lawn pests like cutworms and scarab beetle eggs, and they provide effective control against these lawn and golf course pests.

Even though they're miniscule, they're pretty good at bullying one of our biggest ant bullies: the red imported fire ant. Thief ants are almost three times smaller than the smallest fire ant. Like the James gang, they rely on their cunning and strength in numbers to beat up and eat any upstart fire ant colony making camp in their territory. In fact, fire ants can't establish nests in areas where thief ants roam.

Being tiny has its advantages. Because thief ants nest underground and out of sight, they are one of the few ant species who can weather the havoc wreaked by other nasty invaders like Argentine

ants and yellow crazy ants. When other ant species get kicked out of town, thief ants hold their ground.

Jesse James's shoot-'em-ups and looting sprees came to an abrupt end when he met the wrong end of Robert Ford's pistol. Fortunately, thief ants survive even the toughest ant assassins. They beat up fire ants and outwit Argentine and yellow crazy ants. Unlike Jesse, who caused trouble with

Thief ants tend root aphids underground.

the law wherever he went, thief ants contribute to our natural world. They help keep other pieces of nature in check by eating dead insects and aerating the soil with their underground nests. You could even say they are lawmen in their own right, nibbling away at the pests crawling around your lawn. They're tiny but tough, and they're outside your door right now. Despite their name, thief ants live mostly on the good side of the law.

11 HIGH NOON ANT

SPECIES NAME: *Forelius pruinosus*

SIZE: 1.8–2.54 mm (0.07–0.1 in.)

WHERE IT LIVES: Masters of climate, high noon ants can nest just as happily in your kitchen cabinet as they can in the middle of the desert. They prefer grassy or open ground and often nest under rocks in the central valley, the north, central, and south coasts, and the desert regions of California.

WHAT IT EATS: Excellent scavengers, high noon ants will eat meaty foods like dead insects and animals, but they often prefer liquid sweet treats, like those produced by aphids.

The first time I ran into a high noon ant worker, I'll admit I was underwhelmed. I was ant hunting in a grassy park, laying bait to see which ant species lived in this anty jungle. I'd brought along the

perfect enticement: tuna fish mixed with honey. I measured out this ant catnip onto an index card in a tiny spoonful, which I placed on a spot of bare ground under an oak tree. Then I lay in wait to see who would show up.

Before long, many of my old friends came nosing around. A rusty red field ant with speedy long legs was the first to arrive at the party, bending down, legs spread wide like a horse, to drink in the buffet. She was followed by a small flock of odorous house ants, who were chased away by a steady throng of shiny little black ants. A few acrobat ants briefly lurked around the index card's borders considering the feast then returning to their tree, evidently thinking better of it.

As the little black ants began to scatter, a collection of tentative ant workers I didn't recognize loitered in a tidy line on the sidelines. Plain Jane—brownish red and about half the size of an apple seed—these ladies were unremarkable in appearance. Unlike the frantic field ants or the spirited little black ants, they were a bit boring.

Watching these austere, drab ladies as they efficiently carried off the remaining bits of index card bounty, I almost felt sorry for them. Where were the great spines of winnow ants? The gargantuan size of wood ants? The giant noodles of big headed ants? The happy, heart-shaped bottoms of acrobat ants? Unembellished at best, high noon ants (very common ants who only got a common name as a result of the School of Ants study) don't make a knockout first impression.

As I became better acquainted with them, I learned there's more to high noon ants than meets the eye. So, in salute to the lesser known (and maybe less appreciated) high noon ant, I give to you a countdown of what I believe to be her top five, most notable attributes:

Five. High noon ants are masters of climate. They love to nest in open areas and are able to survive just as well in temperate fields and your kitchen or bathroom as they can in deserts.

Four. They smell good. Like odorous house ants, high noon ants have a pleasant aroma when you smash them. Their bottoms are packed with a chemical that smells sweet, almost like something you'd use to clean your counters. But don't be fooled by their bouquet bottoms: This smell-good

chemical is actually an alarm pheromone that attracts their nest-mates, who then form a mob to help their sister in danger.

High noons eat meaty insects and sweet honeydew.

Three. These clever ants have an entrepreneurial spirit. They have figured out a way to trade their bodyguard skills, whether they're looking out for other insects or for plants, for their favorite syrupy treats. Take the catalpa tree, which gets the most out of this ant's fondness for dessert. When catalpa trees put out their leaves, hungry little Jabba the Huts called catalpa caterpillars come to gobble them up. If nobody protects the tree from these fat piggies, catalpa caterpillars can eat every last leaf off the tree, hurting the tree's ability to produce the food it needs to survive. To save themselves from starvation, catalpa trees put out an SOS call, to high noon ants.

As soon as a caterpillar chomps down on a catalpa leaf, the tree oozes nectar along its branches, which attracts high noon ants. The ants don't want anybody messing with their sugar stash, which means the unsuspecting caterpillars are out of luck. Other plants, like wild cotton plants, similarly benefit from high noon ants' fierce love of sugar.

But high noon ants have no special loyalty to plants. When caterpillars offer to pay them in sugary treats, high noon ants are quick to take the job. Sometimes, they visit and protect the endangered Miami blue butterfly caterpillars, who also offer them a sweet reward for their efforts.

Two. They have the absolute *worst* party etiquette. To keep other diners off the buffet, high noon ants go to those ants' nest entrances in large numbers. Instead of dropping off an invite, they spray bug repellent on the nest entrance, driving the nest's inhabitants deep down. Then, they block up the entrances so those other ants can't escape. Voila! All you can eat is ready and waiting just for the high noon ant.

One. High noon ants can dance! Whenever these ladies are faced with conflict or danger, they try to appease their opponents with a little jig. They shake their bodies around like they're doing the jitterbug and then point their bottoms in each other's faces. This is how they size each other up. Performing these dance moves might not make us the belles of our balls, but they save high noon ants from a lot of cuts and bruises.

So what if they don't have giant spines or huge heads? So what if they don't have cool-shaped bodies? As it turns out, while high noon ants may *look* polite and buttoned up, they're anything but run-of-the-mill workers. With their sweets-loving, booty-shaking, party-crashing ways, high noon ants are a great reminder that we should never judge an ant at first glance.

12 *LASIUS* ANT

SPECIES NAMES: *Lasius niger, L. neoniger,* and *L. alienus*

AKA: small black ant, cornfield ant

SIZE: 3.0–4.2 mm (0.12–0.17 in.)

WHERE IT LIVES: *Lasius* ants prefer open spaces and set up their anthills in grassy areas like golf courses and traffic medians. They sometimes nest under stones or in logs. *Lasius alienus* are heavily distributed along the northern half of the West Coast, are found in the Southwest, and are sprinkled along the East Coast; *L. neoniger* are found throughout the United States; and *L. niger* are found along the West Coast and are heavily distributed in Utah and New Mexico.

WHAT IT EATS: Aphid experts, *Lasius* ants tend aphids like cattle, milking them for honeydew and sometimes killing them for a big aphid steak.

Before I met my husband, an electrical engineer, the only thing I knew for sure that engineers did was drive trains. Sure, I knew tons of people who said they were engineers of one kind or another, but

once I found out they weren't wearing cool hats guiding thousands of tons of steel hurtling at lightning speeds down railroad tracks, my brain would glaze over and I'd lose interest in their profession. "You mean you have no giant horn to blare at passersby as you cross over city streets and along the countryside? BOR-ing!" I used to think.

Then I met the man who would become my husband. As he talked about his job, I learned that many different kinds of engineers work to keep our world safe and running smoothly. Civil engineers, for example, prevent tragedy by designing safe roads, buildings, and bridges. Electrical engineers plan the circulatory system of wires and currents coursing through our cities. Aerospace engineers launch us into the sky and even into outer space. Many more types of engineers tool away behind the scenes, perfecting our lives without our noticing.

In nature, some ant species work as engineers, shaping and refining the environment, building connections, repairing bonds. *Lasius* ants are one group renowned for being superior soil engineers. A little larger than a sesame seed, *Lasius* (LAY-see-us) ants look more like regular ol' ants than the movers and the shakers of nature. They resemble odorous house ants with slightly larger (and fuzzier) behinds and are generally darkish brown to deep brownish black in color. While odorous house ants move like troopers marching in line to and from their favorite foods, *Lasius* ants move more like my favorite Aunt Ann: deftly, but with a bit of a waddle, probably owing a bit to those fuller fannies.

Lasius nests often look like mini-volcanos erupting from the grass.

Three *Lasius* species make the most common list: the small black ant (*Lasius niger*), the cornfield ant (*Lasius neoniger*), and *Lasius alienus* (no common name yet). All *Lasius* species have only one queen per colony and prefer to nest in soil in open areas or under logs and stones. Oftentimes, their nests look like little volcanoes popping up across grassy areas, and, since one nest can have many entrances, they can look like an erupting mountain range spreading out across the landscape.

Lasius ants are often the most abundant ant species on golf courses, happily setting up shop in the expansive open habitat that seems tailor-made just for them. These ants' abodes get them into trouble with golfers, who would prefer their putting greens smooth and free of ant-made speed bumps. But researchers studying *Lasius* ants have shown that when exterminators try to smooth the greens by poisoning *Lasius* ants, more destructive golf course pests like Japanese beetle larvae and cutworms thrive.

In addition to the rolling green, *Lasius* ants love golf courses because they love sugar. Yes, these pristine courses may be free of half-eaten Snickers bars, but *Lasius* ants have their eyes on a different type of sugary treat: chubby, sugar-making insects called root aphids who slurp away at underground grass roots. *Lasius* ants show off their extra-engineering talents as farmers by protecting these root aphids from predators with little mouth-built sheds, all the while milking them for their sugar.

Lasius ants are "aphid experts." They can climb up a tree and tell which aphids like ants and which don't just by sniffing with their antennae. No dummies when it comes to sugar, they prefer the sweetest sugar around. They'll walk around trees sniffing aphid butts until they find the species that produces the sweetest sugar. Whoever wins the sweetest award gets protected and tended by the ants. They're so good at kicking out predators that a ladybug will avoid laying eggs in any area where she even catches a whiff of *Lasius* ants.

Just as some farmers need to kill part of their cow herd to eat meat, *Lasius* ants need to kill some of their aphid herd for protein.

Watch ant nests long enough, and you're sure to see a resident emerge, like this *Lasius* ant.

How can you choose your best beef from all of your best-producing Bessies? If you're a *Lasius* ant, you follow your nose. *Lasius* ants eat aphids from their herd that they or their sisters haven't tended. They can tell who's been milked by sniffing the aphids to see if they've been touched by a member of the colony.

But enough about sugar. Back to the engineering. *Lasius* ants engineer soil. To understand how they help develop our dirt, we first need to understand a little bit more about good ol' terra firma, which props us up right this second, whether we think about it or not. Even though dirt might seem as dead as dead gets, healthy soil actually lives and breathes just like we do. As dead plants and animals decompose, they release nutrients and gasses. Microorganisms then scoot around gobbling up some nutrients and turning them into other nutrients. Animals like earthworms push the dirt about, letting air flow through, speeding up biological processes that increase life-promoting properties within the soil. When nobody contributes to these processes, plants and animals can't survive because the soil gets compacted and hard, with nutrients concentrated in some places and scarce in others.

Lasius ants are among the first species to set up their tents in disturbed areas, and from the moment they move in, they start turning the soil and making the ground ready for life. They burrow their tunnels into the soil, tripling soil respiration, the soil's ability to "breathe," as compared to non-*Lasius* areas. They do such a good job at aeration that more insects and spiders move into those *Lasius*-affected, loosely compacted, fresh-air areas, which has a cascade of positive effects on the ecosystem around them. They carry dead insects and sweet treats from one place to another, spreading out some soil nutrients and concentrating others, which increases all kinds of chemical and biological processes. They even likely help fix nitrogen, an important chemical for healthy plant growth. Just like the engineers who build our world and help it flow, *Lasius* ants churn and build wherever they live.

When was the last time you rode over a bridge or walked into a building and thought, "Gosh, isn't it nice? I'm not afraid this building/bridge will collapse on me"? Probably never. We don't think about it because we don't have to. Thanks to our engineers, things work.

The same goes for soil in our natural areas. We seem to pay attention to our soil only when it stops working well—when plants die, animals move out, and the soil actually becomes stone dead. Our *Lasius* engineers, like my husband and the engineers that reinforce and shape our human-made world, don't seem to mind that we don't notice. They don't need us to shower them with thanks, but they do need us to give them the space and resources they need to do their best work. Keep your eyes peeled for volcanoes on the putting green or in your city's medians. See if you can spot a sesame seed waddling out. Give it a salute as it passes by, conducting its business of building and bettering our world.

13 FIELD ANT

SPECIES NAMES: *Formica pallidefulva, F. incerta,* and *F. subsericea*

SIZE: 5–10.2 mm (0.2–0.4 in.)

WHERE IT LIVES: *Formica pallidefulva* and *F. incerta* usually build mounds in the open, away from trees, while *F. subsericea* generally build their nests against trees, under rocks, or in logs. *F. pallidefulva* are found in South Dakota, along the Southeast and Mid-Atlantic regions, and in the Southwest. *F. incerta* are found along the Mid-Atlantic. *F. subsericea* are found along the northern Midwest region and the eastern half of the United States.

WHAT IT EATS: More buffet goers than picky eaters, field ants love sugar such as aphid honeydew, soft-bodied insects like caterpillars, and seed husks.

Formica ants, usually called field ants, are among the United States' largest and most common ants. Found spanning the states in all directions, three species make the most common list: *Formica pallidefulva* and *F. incerta*, both rusty-to-deep-red beauties, and *F. sub-*

sericea, black lovelies with stripes of sparse golden hairs across their rumps. Most field ants pass their days contentedly building their shallow, low-mound nests near rocks and trees, blissfully unaware of a dark underworld in their midst, a world of violence, slavery, mistaken identity, and poop shields.

About the size of one and a half pencil erasers, field ants' long, dexterous legs extend from their thoraxes, and their large black eyes rest right behind their always-moving elbowed antennae. You can reliably identify an ant as a field ant if it's a large ant, yellowish, reddish, black, or red with a brown or black rump. Many people confuse field ants with carpenter ants, neither of which can hurt you. If you'd like to tell if you have a field ant, gently nab the ant in question and check out its thorax, the middle section of the ant where all the legs attach. If its thorax consists of two lumps, you have a field ant. If it has one big hump, you're holding a carpenter ant. What a way to impress your friends!

Field ants have large eyes because they usually move around during the day and rely on sight more than some other ant species. They use those big eyes to help them see landmarks as they scurry to and from food. Like many ant species, field ants love tending aphids and scale insects for their sugary emissions, but they also help disperse plants by toting seeds around the forest, snacking on the husks and discarding the rest. They also enjoy wolfing down other insects whenever they get the chance.

Like other ant species, field ants tend aphids like cattle, milking them for sweet honeydew and occasionally gobbling a few like steaks.

Nice Outfit, Mr. Beetle

Field ants prefer to eat soft-bodied insects like caterpillars and beetle larvae, and this predatory tendency helps keep our trees happy. One of the northeastern United States' most dangerous forest pests is the gypsy moth. Thanks to their huge appetites, gypsy moth caterpillars have gobbled up more than 80 million acres of our northeastern forests in the past 40 years. When they scarf down all the leaves in the forest, trees die, causing millions of dollars' worth of damage. Fortunately, field ants love those plump little leaf munchers. They help reduce the damage and spread of gypsy moths by eating every caterpillar they can find.

Sometimes their partiality for pudgy little insects lands field ants in unusual situations. Many baby beetles (often called grubs) fit the mold for a perfect field ant meal. Slow, soft, and chubby, beetle grubs don't stand a chance when hungry field ants stumble across them while foraging. To ward off potential beetle slayers, many beetle species, like tortoise beetles, rely on an inventive solution: poop shields.

Here's how it works: Some plants in our forests and across our cities have certain "stop eating me!" chemicals in their leaves, called deterrents. When most insects bite into a leaf and smell the deterrents, they get as far away as possible. Not our resourceful beetle grubs. They eat as much of these stinky leaves as they can, pooping stinky leaf poop all over the place. Then they gather up  the poop and stick it on their bodies, making a force field of stink that follows them wherever they go. Field ants catching a whiff of these otherwise tasty tidbits run in the opposite direction of our little Pigpens. If you feed these baby beetles nonstinky plants, they still make a poop force field, but because it contains no deterrents, field ants will ignore the BM blanket and eat them right on up.

Slaving Away

It may seem like all fun and games for field ants, frolicking across our forests, lawns, and traffic medians, grocery shopping and building their houses. But field ants have a wicked foe prowling those same forests, lawns, and traffic medians, combing the grass for field ant nests: Amazon ants. Amazon ants look a lot like field ants: same size, similar color, same big eyes, similar camel-humpy back. Amazon ants and field ants look so similar they could pass as the same species, almost: Amazon ants have dagger-sharp, sickle-shaped jaws. Their jaws are so pointy they can't take care of tender babies—any attempt at carrying or feeding could result in a fatal stab wound to their young.

So Amazon ants came up with a solution: They raid field ant nests, frighten adults into submission with poofs of chemicals, snatch up hearty pupae in those jaws, and scurry back to their nests. Now, we

Amazon ants carry slaves, unlucky field ant pupae.

remember from the ant's life cycle that baby ants take a lot of food, but once those ants pupate, they don't eat at all. They just sit there helpless in their nests and wait to turn into adults. By stealing pupae, Amazon ants basically snatch up soon-to-be adult workers that require no maintenance in the meantime.

Once in the raiders' nest, the field ant pupae start to pick up the smells in the nest. Ants tell one another apart by smell. If a field ant starts to smell like an Amazon ant, she'll start to think of herself as an Amazon ant. When she emerges as an adult, she will do the tasks to help the colony that she would have done in her real mother's nest: gathering food, building the nest, raising babies, taking care of the queen. She usually will have no idea that she's a slave, helping her enemies to grow so they can raid more field ant nests.

Each summer, poor field ants are enslaved up and down the United States, from the forest near my North Carolina house to the parks of busy Long Island, New York. But you and I can still spot those lucky enough to have escaped the dagger-jaws of the Amazon ants. They run along our tree trunks and across our sidewalks, planting seeds, snagging bugs, turning soil. We can look for their double humps and drop them a snack like a piece of a cookie or some juice from our juice boxes and see if they eat it. We can give them advice, telling them to stay away from poison poop and to keep their big eyes peeled for slave makers. They can give us advice, too. They can tell us never to underestimate the power of small things, to be mindful of the good they can do. They can tell us that every animal has a complicated story, a life of adventure and trials that unfolds whether or not we humans pay attention. But you and I can pay attention. Field ants are happy to share their story with us.

14 ARGENTINE ANT

SPECIES NAME: *Linepithema humile*

SIZE: 2.2–2.6 mm (0.09–0.1 in.)

WHERE IT LIVES: Argentine ants prefer cities and form loose nests in moist landscaping materials such as mulch, under trees or walkway bricks, in potting soil, or under rocks. In California, they are found on the North, Central, and South Coasts and in the Central Valley.

WHAT IT EATS: Argentine ants scavenge and devour insects and other small arthropods and human garbage, but they prefer sweet liquids such as the honeydew produced by scales and aphids.

When I first started studying ants, my friend and lab mate Alexei took me to an office park just outside of Raleigh to hunt for ants. Seeing nothing but boring, run-of-the-mill brick office buildings sep-

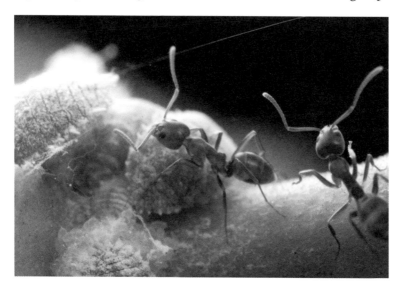

arated by expansive parking lots dotted with sickly trees, I started to think he had misunderstood our mission. Where were the forests with acrobat ants waving their fannies happily along tree trunks? The lush backyards of strangers hiding thief ants, winnow ants, and maybe a fat, grumpy box turtle or two? At the very least, where were the odorous house ant–packed dumpsters behind city restaurants— restaurants that might have an ice cream cone waiting inside just for me? Alexei pulled into one of the parking lots. It looked like all the other parking lots in the park.

"We're here," he said, and we hopped out of the truck.

We wove between parked cars and skipped over empty spaces. He peered behind sedans and pickup trucks, crept around coupes. Meanwhile, I developed a working theory about the strength of his sanity.

Suddenly, Alexei pointed. "Look right there!"

Hugging the curb, a thick braid of dead ants stretched the length of the parking lot as far as I could see in either direction. It twisted and bent in the wind.

Alexei explained we were standing on an Argentine ant (*Linepithema humile*) supercolony boundary line. He said they fight to defend their colony borders every day, which accounts for the huge number of casualties. I picked up a piece of the dead ant rope and looked more closely at these Argentine ants. About the size of two sesame seeds and reddish brown, they didn't look like much. Soon I learned that, despite their looks, I was holding a fistful of some of the most extraordinary ants in the world.

Argentine ants set sail for the United States from Argentina and Brazil in the late 1800s. Stowaways in ships' ballasts, they made port in New Orleans and quickly set up camp. Before long, they fanned out and blanketed the city, nesting in every nook and cranny they could find. They gobbled up everything they could. One researcher in the early 1900s wrote that Argentine ant populations were so humongous, they ate all the bedbugs in town.

Back at home, while their rambling sisters were out gobbling up all the crawfish étouffée they could eat, the native Argentine ants weren't a problem, quietly nesting in floodplains and other moist areas. Today in their native land, they still aren't a problem. Argentine ants, like many other ant species, can tell the difference between ants from their home nest and ants from other nests based on the way they smell. When Argentine ants encounter an ant that smells differently from the way their home nest smells, they attack it. As a result, Argentine ants keep each other in check and their colony sizes stay small.

When Argentine ants hitch a ride to other countries, only a few families make it to establish a home. These families, closely related to each other, smell a lot alike, so the workers refrain from fuss-

This Argentine ant queen will work with other mothers in a super-colony to lay thousands of eggs.

ing and fighting. Instead, they concentrate their energy on fighting other ant species, gathering food, growing... and growing. By keeping the peace with each other and dominating other species, Argentine ant colonies turn into supercolonies. While the supercolonies in my state of North Carolina stretch only a few miles, in California one Argentine ant supercolony spans more than 500 miles. Along the Mediterranean coast, one reaches almost 4,000 miles. These supercolonies contain thousands of nests and queens with millions of workers.

All this buddy-buddy behavior among invading Argentine ants can really wreck their adopted lands. Taken alone, one Argentine ant isn't all that impressive. With no spines or hairs, no impressive size or fancy gasters, they look like regular, run-of-the-mill ants. They can't sting. Squishing an Argentine ant is no problem—they can't even run that fast. When individual Argentine ant workers try to defend themselves against other ants, they rarely stand a chance.

But with millions of sisters getting each other's backs, they can team up and overwhelm any opponent. They handily defeat native

Argentines use teamwork to overcome larger enemies.

ants and scarf down all kinds of insects, mites, and other arthropods, and their appetite for sugar leads them to foster sugar-producing plant pests like aphids and scale insects. Altering the environment on such a dramatic scale can have cascading effects across the whole ecosystem.

When Argentine ants remove other ant species, they take away food sources for some animals, and they eliminate the beneficial work of other ants without filling those positions. For example, by kicking out ants native to California, Argentine ants eliminated coastal horned lizard food, leaving baby lizards to starve and lizard populations to plummet. In South Africa, they give seed-dispersing ants the boot, and the plants that grow from ant-dispersed seeds die out. By protecting honeydew-producing (and also disease-transmitting) plant pests, Argentine ants increase plant diseases in their invaded areas. Their sweet tooth also drives them into honey-bee hives in parts of the country, bothering the bees so much that they abscond for ant-free areas, leaving beekeepers to shake their fists. Left unattended, Argentine ants can empty a whole beeyard of hives.

With so many scouts combing the area for snacks, Argentine ants can locate ant food faster than other ant species. Once at the snack site, they get all their sisters involved to bully any other would-be snackers into staying away.

Protecting the environment from these little raiders can be extremely hard. They have picked up a few tricks from their native land that keep them just ahead of the insect control curve. Because they evolved in floodplain areas, where at any time their home could be underwater, Argentine ants are good at moving from place to place. Their many-queened nests are loose groups of ants under mulch, doormats, pine straw, in home walls—anywhere moist and dark—with a cluster of white brood and eggs. When danger strikes, they pack up the young 'uns and move. As they move, they can split their queens into more nests. Now,

Argentine ants gather honeydew from scale insects.

one nest has become several, and the Argentine ant population has room to grow.

Because of this, people often find Argentine ants breaking into their homes and office buildings, scarfing down their sodas and hanging around their trash cans. We try to knock them back from

our property, unaware that millions of them wait on our borders, ready to move into the empty spots left by their sisters.

Yes, tiny Argentine ants cause all sorts of trouble. Yes, it would probably be best if we sent them all packing back to Argentina, where their natural enemies and cousins would give them a stern talkin' to. But for now, Argentine ants aren't going anywhere. Some researchers are capitalizing on this, watching how Argentine ants run around when they're panicking, to help us understand how people can safely exit dangerous, crowded situations.

After that first day with Alexei in the office park parking lot, watching the dead Argentine ants blow like a long tumbleweed in the wind, I spent many years observing Argentine ants. I watched their brown bodies glimmer in the sun as they trailed up and down maple tree trunks, carrying honeydew in their fat bottoms. I saw pieces of popcorn seem to float across the pavement as these ants carried them back to their nest. I watched them disassemble a whole grasshopper before my eyes. I've uncovered nests all thick with white brood and eggs and watched them carry that whole nest to a new location in minutes. I saw these tiny individuals persist, grow, and dominate by working peacefully together.

15 RED IMPORTED FIRE ANT

~~~~~~

SPECIES NAME: *Solenopsis invicta*

SIZE: 2.54–6.1 mm (0.1–0.24 in.)

WHERE IT LIVES: Red imported fire ants build their mounds in a variety of habitats, including yards, parks, along roadsides and forest margins, and near swamps across the southern United States. They are heavily distributed in the Southeast region but have also been found in the Southwest and in California.

WHAT IT EATS: Red imported fire ants love protein and sugar and will eat anything from dead insects and honeydew to people food.

~~~~~~

Anyone growing up south of Virginia or east of New Mexico has likely experienced the supreme pleasure of jamming a stick into a red imported fire ant mound and watching its angry inhabitants boil

This red imported fire ant waves her venom-tipped stinger.

out from the earth. Fire ants are notorious for their brawly disposition and their pustule-producing stings, but beneath that surface
lies an intricate, well-oiled machine, or perhaps the better word is
superorganism, that's worth a second look (hold your sticks, please).

Fire ants originally hail from Argentina, Brazil, and Paraguay,
but ever since they landed in Mobile, Alabama, in the 1930s, they
have made themselves at home in the United States. Their conspicuous, large mounds burst from the land across the Southeast like mini
volcanoes, dotting open ground, roadsides, agricultural fields, and
residential areas. Fire ants fill their mounds with galleries and tunnels for storing food, raising young, and just hanging out being ants.

Fire ant mounds also have their own air-conditioning systems.
Because these mounds extend deep into the earth, the ants can move
their colony down to cooler ground when it's hot outside. When
temperatures drop, they pull the colony up closer to the surface to
take advantage of the sun's warming rays. Fire ant colonies can grow
to be large. Some include as many as 200,000 workers, roughly the
size of the inner core of my own home city, Raleigh, North Carolina.

In order to make a nest large enough for so many respiring, moving, working, eating bodies, they have a lot of soil to haul. The dirt
on the surface of some fire ant mounds (most of which corresponds
to the holes they have dug below) can fill 10 gallon-size milk jugs.
Meanwhile, below the surface, their underground foraging tunnels
can radiate out 100 feet in all directions, allowing them to reach a

cool drink of water or food even when the air above ground is too cold or (just as often in the South) too hot.

All of that soil turning aerates the ground and helps plants grow, but it gets them—and us—into trouble. Agricultural fields with plantings like soybeans and corn provide everything fire ants need in one place, a kind of agricultural Walmart. The soft, tilled earth makes a perfect nesting ground. Pest insects cover the plants—easy prey for fire ants to pick up and take back home. Fewer native ant species roam these fields, offering the fire ants less competition. Farms are where, to these ants, the livin' is easy. Unfortunately, fire ants don't realize farmers plant fields for people, not fire ants. Their mounds damage harvest machinery, and the ants nibble

plant roots, resulting in lower crop production. In addition, fire ants encourage pest insects like aphids to grow so the ants can then snack on honeydew, the sweet substance aphids produce. Plus, they sting anyone unlucky enough to cross their threshold.

When a fire ant stings you, she first grabs your skin with her jaws. Once she has a tight grip, she jabs her stinger, located at the tip of her abdomen, into your skin as many times as she can before you knock her off. Her stinger is a tiny poisoned spear. Each time her stinger makes contact with you, she injects a small amount of toxic poison. This poison causes itching, oozing pustules on most people, but for some unlucky victims, the stings trigger an extreme allergic reaction, anaphylaxis.

TOP:
Fire ants build conspicuous mounds.

BOTTOM:
A fire ant bites to hold on while she stings.

For their friends, their stingers and jaws do more than just fuss and fight. In addition to restraining prey and intimidating enemies, fire ants' combination of gripping, pinching mandibles and precise stingers actually work most often as versatile tools that help the workers lay down chemical trails, handy for finding their way back home from food resources.

Everything I've told you so far is about how fire ants live in the places where they have invaded (or rather, where we have brought them). Back home in their native Argentina, fire ants nest near frequently flooded river beds. Most ordinary ant species would drown in these floods, but not our fire ant. (Note: as a Southerner, despite not always liking fire ants, I do think of them, a bit, as "mine.") When the water rises, these ladies hook their legs together to make massive living rafts for their entire colony to float along and ride out the flood. Workers take turns riding the waves so no ant is left under water for too long. Covered in small, water-repellent hairs, fire ant bodies stay coated in a silvery sheen of air even if you try to dunk them underwater.

Young, stick-toting naturalists aren't the only observers captivated by fire ants' curious lifestyle. In fact, the fire ant is the most studied of all ant species; scientists have spent years tracking its behavior and ecology. Scientists and citizens alike try to prevent fire ants from marching forward across the United States. We use an array of poisons and home remedies like boiling salt water. We release enemies like fire ant–eating flies. We make laws to prevent moving them from state to state. But, despite these efforts, fire ants remain, turning soil, laying trails, tending aphids, making rafts. Fire ants live up to their scientific name—*invicta*, or "unconquered."

16 CRAZY ANT

SPECIES NAMES: *Nylanderia flavipes, N. terricola, N. fulva*

SIZE: 1.0–4.0 mm (.04–0.2 in.)

WHERE IT LIVES: Crazy ants aren't choosy about where to nest and can make their homes under trees or in your potted plants. *Nylanderia flavipes* are found sprinkled along the Mid-Atlantic and Northeast regions; *N. terricola* are found in the Southwest region; and *N. fulva* are found in South America and the Caribbean.

WHAT IT EATS: Nuts for sugar, crazy ants prefer sweet syrups produced by aphids but will snack on human garbage or scavenge for insects if they get the chance.

Crazy relatives. There's at least one in every family. I have several in mine. Take Aunt Nee Nee, who brings my grandfather along to all our family events. My grandfather died 15 years ago. His ashes reside in a wooden box the size of two dictionaries stacked on top

If you look closely, you can see the crazy ant's crazy hairdo.

of each other. At my wedding, she propped him in the choir loft, "so he can see," and had the photographer do a photo session with him. Or Uncle George, who went missing for two weeks. Just when everybody thought he was dead, he came rolling into town in a pink Cadillac with a live monkey strapped in the passenger seat. Or Aunt Ann, who... well, you get the picture.

Ants have a lot of crazy relatives, too. Most members of the genus *Nylanderia* even get the common name "crazy ant." Some of them deserve it.

They even have crazy little hairdos. If you're like me and you have a special place in your heart for all things fuzzy, crazy ants are the ants for you. Ranging in color from pale yellow to black and a little bigger than a sesame seed, crazy ants have spiky hairs covering their entire bodies that make them look like baby birds or old men's heads. Either way: a-dorable. They'll nest in any nook or cranny they can find, squeezing into potting soil or snuggling up to trees in medians. They eat pretty much anything, too. From honeydew to small insects to our trash, crazy ants aren't picky. Three species make the most common list: *Nylanderia flavipes* (the yellow-footed ant); *N. terricola*; and, the ant equivalent of Aunt Annie Kate (the craziest), *N. fulva* (the Raspberry crazy ant).

Crazy ants get their common name from the way they run around like a house afire while they're foraging. Most ants seem to move about in orderly lines or careful steps, but crazy ants have a wiggle-waggle way of running as they go to and from food. Between their sparse fur coats and their zany walk, you shouldn't have a hard time telling these nutty ants from the others running around your city park.

Their easygoing ways and catholic diets make crazy ants excellent party crashers. Because they can nest anywhere and eat anything, crazy ants have no problem moving into new environments. When animals like crazy ants aren't fussy about what they need, they can expand their empires into new locations more easily than their choosier chums. Unfortunately, this means these easy-to-please insects often become invasive pests.

Yellow-footed crazy ants originally came from Asia but have made themselves right at home across the eastern and midwestern United States, from Washington, DC, to Philadelphia and New York City, spreading west to Cleveland, Ohio. Once they move in, they gobble up all the food and make lots of babies, eventually taking resources from other ant species and amassing huge worker populations. By now, we all know that our native ants do many good deeds for our environment, from engineering the soil to keeping tree canopies healthy to planting our wild herb seeds. Suppose a crazy ant like a yellow-footed ant moved in and took all of the homesteads from our native ants or ate all their food? What would happen to our homegrown heroes? And what about the jobs they do? Many times, invasive species like yellow-footed ants spell trouble for our natural world.

But yellow-footed crazy ants are just the tip of the iceberg of crazy in *Nylanderia* world. Raspberry crazy ants, named for Tom Raspberry, the exterminator who discovered them, take outrageous ant behavior to a whole new level.

Raspberry crazy ants originally came from Brazil and Argentina but made a homestead in Houston, Texas, a few years ago. Just like

Crazy ants can have tremendous populations and can march into people's homes.

yellow-footed ants, they began to scarf down all the ant food they could find, turning that ant food into ant babies. Lots and lots of ant babies. In some places, Raspberry crazy ants became so numerous they overran human structures.

They can reach such huge populations that they become more than just a nuisance for humans; they become real trouble. Raspberry crazy ants, like their crazy ant cousins, nest in all kinds of crannies, and electrical boxes make perfect crannies for them. If an ant worker gets zapped in an electrical box, she'll release a "Danger!" odor, called an alarm pheromone, which alerts her sisters that she's in need of assistance. Her sisters will pour in en masse and flood the box, shorting out electrical equipment. The more ants that get zapped, the bigger the danger odor and the more ants pour in. It's not unusual for people in the southwestern United States to open electrical boxes and find them packed with tens of thousands of electrocuted ants.

Crazy ants also get a case of the hangries. When these ladies get low on sugar, they become super aggressive and start swinging punches at anybody within reach. Like all *Nylanderia*, these bizarre trespassers can't sting, but they build up such tremendous populations that when they come to town, they can even wipe out the ornery red imported fire ants.

While we might not always appreciate our crazy relatives (Uncle George's monkey was a biter), they certainly do make our lives more interesting. Sometimes, they might take over the party. Even so, they're fun to watch, and you can always count on them to show up.

17 TRAP-JAW ANT

SPECIES NAME: *Strumigenys* spp.

SIZE: <2.54 mm (<0.1 in.)

WHERE IT LIVES: Tiny trap-jaw ants live in rotting logs, under stones, and in the soil—anywhere near their favorite meal: collembolans. *Strumigenys rostrata* are primarily found throughout the Southeast and Mid-Atlantic regions.

WHAT IT EATS: Stealthy-but-blind hunters, trap-jaw ants stalk between grains of dirt for little soil dwellers like collembolans, mites, and termites. But their favorite meal is collembolans.

It's dark. You're surrounded by stones and boulders. Huge green vines thrust and spiral outward from the earth. Woody mountains knot the terrain all around you. Springtails—large, soft creatures—peacefully push the rocks with their antennated foreheads and ten-

Trap-jaw ants are so small that, to them, moss fronds are a forest.

der, fingery legs, foraging like cattle for decaying plants under the rubble. Somewhere behind the boulders, a long-toothed predator lurks.

In the shadows you can just make out its dagger-jaws, gaping and ready for its next meal. You see its narrow, skull-like cranium pitted and dotted with sparse hairs, its slender, knobby, stealthy killer's body, grooved and hairy, tipped at the end with its deadly stinger.

You watch in the darkness as it tip-tip-taps its own antennae one way and then the next, mandibles locked open, using scent to stalk the soft-bodied creatures around. The predator moves cautiously, deliberately, getting closer, closer still to its unsuspecting meal. Its antennae lightly sweep its prey; its jaws sneakily surround the victim's body. When the prey moves, even slightly, against those waiting jaws: CRACK! In less than two-thousandths of one second, the hunter's spiked mandibles snap shut around its quarry, impaling the soft flesh. The poor animal tries weakly to struggle, but the predator lifts it high in the air and stabs it with a venomous sting, paralyzing it for the journey home to the predator's nest, where its sisters will disassemble the meal for a family feast.

This resident assassin lurks in the darkness all around you, but you don't need to be afraid. The *Strumigenys* ant, our sickle-toothed hunter, measures less than one-tenth of an inch—nearly as tiny as the grains of dirt between which she navigates. From our human-size point of view, she is almost too small to see without a magni-

fying glass or microscope, so small you could pack half a colony of her kind into a buttonhole. And while she may strike terror in the tiny hearts of soil dwellers, to me she looks like an unearthed jewel, something rare and adorned and stunning.

Strumigenys, the tiniest type of trap-jaw ant, whose name literally means "tumor jaw" for its large, powerful mandibles, is common in moist soil, rotting logs, and under rocks across the United States. However, many people don't realize ants like these live underfoot. Because of their diminutive size and their soil-dwelling lifestyle, these ladies can be hard to spot as they cruise the underworld for springtails. Two species make the most common list, *S. talpa* and *S. pulchella*, but many more *Strumigenys* species stalk the soil around us. Although some *Strumigenys* will settle for juicy mites or other soft-bodied soil dwellers, most prefer a springtail supper.

Springtails, also called collembolans, are like the Holstein cows of the undergrowth, enormously abundant and fantastically nutritious. Not quite insects, springtails look like insects that haven't quite been fully formed yet. Some are round and fat and almost could be mistaken for miniature aphids, while others have longer bodies with peaceful black eyes and long, bouncy antennae. All springtails, no matter how slender or plump or bald or hairy, have long tails, called furculae, which they tuck under their bodies. When a springtail senses danger, it will pop its furcula down like a spring, striking it against the ground, propelling its body high in the air—sometimes to a distance almost 100 times its own body length. Because of this emergency escape route, *Strumigenys* need to be super stealthy when on the hunt for springtails, and these ants adopt a distinctive, creeping, sneaking gait that makes them seem like they're always up to something.

Nearly blind, *Strumigenys* ants rely on their sense of smell to locate and capture springtails. Because springtails can't see too well either, they often accidentally crawl right over unsuspecting *Strumigenys* backs without the ants noticing they've missed a good meal. The ants also consume mites, tiny termites, and other minute creatures crawling through the dirt. When such a creature happens by and grazes the trigger hairs between a *Strumigenys*'s jaws, the jaws snap shut. Sometimes, a *Strumigenys* ant will sneak her jaws around a potential meal, only to find the meal doesn't bump into those trigger hairs. If this happens, she'll wait for a few minutes in case the bug wants to wake up. If it continues to sit still, she will tap its back with her antennae to try to wake it up to an unpleasant surprise.

Tiny trap-jaws live in correspondingly tiny nests excavated out of rotted wood or damp soil, sometimes under stones or leaves. They usually have between 25 and 100 workers crawling through their nests. We don't often see them if we're looking for ants on the pavement or our front porches, but they're there, wandering through the soil and leaf litter. We can find them by stealing a bit of leaves and

soil and bringing it home to the lab or by digging a pitfall trap and waiting to see who falls in.

Strumigenys ants will wow you when you do meet them. They look almost impossible, with their hairs and pits and pivots and jaws. They make me excited to look at ordinary logs and twigs in my yard, to imagine the hunt and capture, the alien-like bodies moving grains of dirt like mountains while I sip my tea.

18 ACROBAT ANT

SPECIES NAMES: *Crematogaster ashmeadi, C. lineolata,* and *C. cerasi*

SIZE: 2.6–4.4 mm (0.1–0.18 in.)

WHERE IT LIVES: Most often, you will find acrobat ants nesting under bark in standing trees, on the forest floor, or in rotting wood, but sometimes they wander into our homes, snuggling their nests in tight spots like between shingles and in the walls. *Crematogaster ashmeadi* live in the southern United States. *C. lineolata* are primarily found in the Midwest, Mid-Atlantic, and Northeast regions and are also sprinkled in the southern United States, as far west as Arizona. *C. cerasi* are heavily distributed in the Midwest and are also found along the Southwest, Southeast, Mid-Atlantic, and Northeast regions.

WHAT IT EATS: Primarily sugar lovers, acrobat ants sometimes take a break from lapping honeydew off aphids' rear ends to forage on protein such as dead insects.

One summer, I traveled to a remote North Carolina island for a research project. The project required that I crawl under and around people's homes looking for ants. While walking to one home, I accumulated a following of local ducks that waddled behind me, waggling their bills in the hope of getting food and quacking reproachfully when none appeared. I hate to disappoint, so I snuck into a local's backyard and dumped out my supplies, looking for a duck-suitable snack. As I rifled through my bug-collecting equipment, a man came out of the house.

"Hello, ducks ducks ducks!" he said, and the ducks happily abandoned me for their old friend. Thinking I was caught trespassing, I shoved my equipment back in my bucket and hurried to introduce the man to this potential thief/weirdo lurking around his backyard. He told me he could hear me coming but couldn't see me; he was blind.

"I'm looking for ants," I explained.

"Ants?" he asked. "I've got acrobat ants! Come see!"

The ducks and I followed him. He felt the way off his back porch, running his rough hands along the brick walls of his house, around the corner, and pushed his body behind a hedge. He pulled back branches from a wax myrtle tree and revealed a pipe leading into his house. On that pipe? A parade of acrobat ants, their little heart-shaped fannies waving in the sun!

I tried to imagine how he could find this tiny treasure so deeply hidden.

"How in the world could you tell these were acrobat ants?" I asked.

"Because," he said, and he slammed his hand down on the pipe, smashing a couple of workers. When he lifted his hand, I watched the stunned workers stumble about, smoothing their crumpled legs and antennae, gradually going back to work. "You just can't squish the jimdurn things."

He was right; acrobat ants seem to defy squishing.

Acrobat ants are a gift, a joy, and you can find them almost anywhere you'd imagine in the United States, from swamps, deserts, and forests to your kitchen cabinet. Three species of acrobat ants make the most common US ant species list: *Crematogaster ashmeadi*, *C. lineolata*, and *C. cerasi*. These species can be hard to tell apart just by looking at them. About half the size of an apple seed, they range in color from rusty with dark brown/black abdomens to a deep reddish-black all over.

Even so, you can tell acrobat ants from other types of ants by their heart-shaped bottoms, or gasters. They trail in happy lines to and from food. When disturbed, acrobat ants halt and wave these hearts in the air like proud flag bearers in a pageant.

Acrobat ants tend their queen with great care.

It's hard to imagine how acrobat ants are among the most abundant ants in forests and homes, considering what a fragile enterprise colony-founding is for them. Imagine a big forest where acrobat ants might live. Picture all the towering trees, with their seemingly infinite number of branches, stems, and leaves, jutting out against the sky. Now picture one tiny ant, a newly mated queen, slightly longer than an apple seed, embarking alone for the journey of her life.

All kinds of animals like spiders, mice, beetles, and birds would love to snack on our queen, and the forest trembles with life as these predators peek and poke about, looking for a treat. Our queen, our apple seed, keeps her course, searching

Acrobat ants hold their gasters high like flags.

the branches for an abandoned beetle or termite gallery to make her new home. When she finds one, she settles in, laying eggs that will become her empire.

For every 100 acrobat ant queens that journey to find a new home, fewer than eight (and usually much fewer than eight) survive to form a colony. Once formed, a colony can live 10 to 15 years and may have anywhere from a few to several thousand workers crawling across the branches, eating everything from nectar to other insects.

Those workers help keep forests healthy and balanced. Acrobat ants help protect or sustain at least two endangered species: the Miami blue butterfly and the red cockaded woodpecker. In exchange for a sweet substance produced by Miami blue caterpillars, acrobat ants feistily fend off would-be butterfly poachers like birds and other ants. They also are the red cockaded woodpecker's primary diet. Wiping out acrobat ants could have a domino effect across the forest, with other species falling down in turn.

Consummate hosts, acrobat ants often harvest clytrine leaf beetle eggs from leaves and, without eating them, bring them into their nests, where the eggs hatch in a predator-free environment. Another ant-loving beetle, *Fustiger knausii*, spends most of its life hanging out in acrobat ant nests, relaxing with the brood and riding around on workers' backs. They groom the ants and might get food by entic-

ing workers to spit up snacks for them to eat! Just what the ants get for their hard work is less clear.

Like my duck-loving friend on the island, you'll find acrobat ants parading around your kitchen or, true to their name, tightroping across your clothesline. Don't be afraid of them! They aren't dirty and they won't hurt you. Many of us commonly encounter acrobat ants

and don't realize it. That's because many of us, unlike my friend, choose to be blind, to ignore these marvels of life as they shiver all around us. Maybe you, like my friend, can take a break to experience the pageantry of the happy procession before you. To enjoy the sensation of those cheery bottoms waving in the air on their way to work. To thank them for the job they do. Just try not to squish them.

Acrobat ants play host to beetles and crickets.

FREQUENTLY ASKED QUESTIONS

Ant researchers often get asked some really interesting questions. For example, what do you do when you get ants in your pants? (Jump around and squeal, obviously!) We like that people pay attention to what's living around them and want to know more. I asked some of my favorite ant researchers what they most often are asked about ants. Here are the answers to some of our most commonly asked ant questions.

What is the biggest ant?

The dinosaur ant, *Dinoponera gigantea*, boasts the largest workers in the world, measuring a little over an inch to more than 1.5 inches long. These whoppers live in South America. But the contest is a close one. At 1.2 inches long, both the Southeast Asian giant forest ant, *Camponotus gigas*, a close relative of our US carpenter ant, and the bullet ant, *Paraponera clavata*, from Central and South America, give dinosaur ants a run for their money. In the United States, carpenter ants can measure up to a half-inch long and probably hold the distinction of the largest ants in the country.

What is the fastest ant?

No one has calculated which ant species runs the fastest, though you might judge your local species for yourself in your backyard with some ant bait and a stopwatch. Still, some ants are known to have some pretty quick moves. One type of trap-jaw ant, *Odontomachus* spp., have jaws that shut at lightning speeds—up to 145 miles per hour. They use these quick reflexes to chomp down on prey and threats alike, but they also use their snappy mouthparts to help them battle intruders or quickly exit any scene. When danger rears its ugly head, a trap-jaw ant can either bounce the botherer off its snapping jaws, tossing the intruder aside, or bounce itself off the intruder, launching itself far from danger. Trap-jaw ants can also snap their jaws down on the ground to catapult themselves into the air and away from danger.

How do I find ants?

As you probably already know, ants are super easy to find. They're all around you! It's best to look for ants on warmer days with low wind speeds, but you can look for ants any time you want.

One of the best ways to find them is to coax them into coming out into the open by offering them a snack. Some ant species love sugar while others prefer protein, so read a little about which species you're trying to find before preparing a meal for them. If you just want to see who's around, you can crumble some pecan cookies onto an index card, a piece of paper or tin foil, or any other flattish surface and wait to see who comes to the party. Pecan cookies offer protein and sugar at the same time. Many ant researchers also use canned tuna fish in oil to find their protein lovers and honey or jelly, like apple jelly, mixed with a tiny bit of water for their sugar lovers. Some researchers mix them both together to make a stinky salad that equates to many ants' dream meal.

While you're watching your ant baits, see if you can identify which species show up. Watch to see how they interact at the baits. Some species bicker with each other, while others ignore everybody completely and stick to the task at hand. Some species lay chemical trails back to their nest to get their sisters excited about this new grocery store in town, while others will carry as much home as they can in their mandibles. Still others will run away, only to return carrying their ready-to-help sisters, whom they plop down on the food. You can follow all these ants back to their nests if you have keen eyes.

Another way to look for ants is to look under things. Logs, flower pots, stones, and mulch are all great objects to peek under or pick through. I like to carefully peel back the bark on rotting wood to see who's living there, and I'm rarely disappointed. Just keep your eyes open for insects and other creatures you might not want to find, like snakes and spiders. They like living under bark and logs as much as ants do. It can be exciting to find them, but sometimes they'll startle you. Remember to replace the log, rock, or bark when you are done peeking, though, so the ants and other small wildlife can return to their business after you amble away.

If you want to take ant hunting to the next level, you can set up pitfall traps. For a pitfall trap, you'll need a little liquid dish detergent (or other liquid soap) mixed with some water (and a little rubbing alcohol, if you have it—or, even better, ethanol) and a small, disposable plastic cup. Dig a cup-size hole in the ground and place your cup in it so the lip of the cup lines up with the top of your hole. Try to get a cup with a small mouth on it so that big stuff won't fall in. Then, fill your cup halfway with your detergent mixture. Leave it in the ground for a day or so, and when you come back, you can see who fell in. You can dump your cup's contents into a white-bottomed tray or, if you don't have that, a clear dish with a white piece of paper under it so you can see everybody. If you leave the cup in the soil too long, the insects you catch will begin to rot and the smell will probably prevent you from wanting to check, so it's best to just leave it out for a day at a time.

If you want to get even deeper into ant-finding, visit your local library and check out a great ant-hunting method book like *Ants: Standard Methods for Measuring and Monitoring Biodiversity*, edited by Donat Agosti, Jonathan Majer, Leeanne Alonso, and Ted Schultz. Books like this will give you all sorts of ideas about how to find the ants you want. Such books are written for serious ant enthusiasts, who, as we all know, include kids, bankers, teachers, and anyone else who wants to seriously study the other societies all around us.

Are ants related to termites?

Ants *are* related to termites in that they're both insects, but they're not closely related to termites, and they have very different lifestyles. While ants have queens who store lots of sperm and lay eggs, termites have kings and queens who mate repeatedly over the course of the colony's life cycle.

Ants also develop by means of "complete metamorphosis," which means the eggs hatch into baby ants (larvae) that look more like fly maggots than they do ants. These larvae must pupate (often by spinning cocoons, though some ants forgo this complexity) and undergo metamorphosis before they are adult workers who look like the ants we see on our sidewalks and trees. Baby termites, on the other hand, hatch from eggs looking like miniature termites.

To tell ants apart from termites just by looking, check their waists and antennae. Ants have narrow "wasp waists," while termites look chubbier around the middle. Ants also have elbowed antennae, whereas termite antennae stick straight out.

Are ants related to wasps?

Ants are more closely related to wasps and bees than they are to termites (or most other insects). Ants, bees, and wasps are all members of the scientific order *Hymenoptera*, which means they share both physical characteristics and common ancestors.

How long do ants live?

We're not sure how long many species of ant workers live. Their lifespans often depend on the season (and whether they get stepped on, licked up by an anteater, chomped by a bird, or taken over by a zombie parasite), but they can live anywhere from a couple of months to several years. Ant researchers typically record how long ants live by how long the colony's queen can survive, since she is the colony's beating heart.

Some ant queens live long, regal lives (if a life consisting mostly of waiting to be fed and laying eggs is regal). Others don't. For example, red imported fire ant queens (*Solenopsis invicta*) can live almost seven years, depending on where they live. Winnow ant (*Aphaenogaster* spp.) queens can live up to 13 years. Some *Lasius* queens have been documented as living nearly 30 years. Other ant queens last less than a season.

Do carpenter ants eat wood?

While carpenter ants gnaw wood to build their nests, they don't actually eat it. Like humans, they just whittle while they work.

What does an ant queen look like?

Some ant queens, like Asian needle ant queens, look a lot like their workers. Others are the same color as their workers but have tremendous gasters, good for egg laying. Ant queens usually have larger eyes and bulkier thoraxes than their workers. The ant queens of some species, such as army ants, are so large and different from their workers that they look like another species entirely.

How many different ant species are there?

So far, we know of about 15,000 ant species roaming the earth, and nearly 1,000 of those species call the United States home. Nearly half of these species have not yet been named and more remain undiscovered, which is to say new ant species may well be lurking in your backyard.

HOW TO KEEP ANTS AT HOME

So You Want to Keep Ants as Pets . . .

Acrobat ants walking tightropes, carpenter ants having some serious conversations—now that you've read about all these creatures crawling around you, you might be tempted to capture some to keep around the house. Watching ants go about their business from the comfort of your home can be fun, especially when they're going about their business in a container and not just wandering around your kitchen counter.

Ant Sleuthing

Before you decide to take the leap into ant ownership, consider becoming a grade-A ant sleuth instead. Ant stalking can offer endless fun, and you'll get the chance to observe many more species

than you would if you kept a species or two in your house. To be a good ant detective, you can either lure the ants to you or hunt them down.

Luring Ants: The Way to an Ant's Heart . . .

When luring ants, consider their favorite meal. Some ants prefer sugary foods (like honeydew), some prefer protein (like arthropods), and some like a little bit of both. Start by reading about your favorite ant species to see what they like best.

Once you know their favorite fare, pick something you have around the house that offers them a taste of what they want. Protein lovers might like a little peanut butter or tuna fish in oil. Those

with a sweet tooth may prefer honey, jam, or sugar water. Cookies are always a good bet. Stay away from "healthy snacks" like broccoli or celery. Ants never seem to care too much for plants (unless those plants have bugs living on them). Feel free to be an ant chef and experiment with different foods as you get to know different ant species.

You'll want a handful of mini platters so you can spread the bait out around your yard. You can use index cards, tin foil squares, or some other flat surface. Place a tiny amount (1/4 teaspoon or less) of your food on each one, and take them outdoors to see what you can find. After you've distributed their bait, wait 20 minutes to an hour and see who shows up.

Watching ants can be very exciting. Some get to the party really early and run away when other ants come around. Others saunter in and frighten away the ants already there. Some ants play well with others and ignore whoever else shows up. Others fight or engage in ritualized displays (where they wave their gasters in the air or open their mandibles or swing their heads to and fro) to show how mean they can be. I've even seen ants show up only to eat the other ants eating the baits!

See how the ants find the baits and how they bring their sisters back to get more food. Do they wander in, or do they make a direct line? Do they come alone, or do they bring reinforcements? If you place two different types of food out for the ants, do they prefer one over the other? You can come up with many questions while luring ants and get a lot of insight into how ants operate just by watching them eat. Be sure not to breathe on them while you watch them. They hate that.

Hunting Ants: No Weapons Necessary

If you've set up a bait, you have a good eye and are quick, you can follow the ants back to their nests. Many of us are not so quick, though, and we need to come up with other ways to hunt ants.

Hunting ants can seem tricky at first, but really it's not so hard. In the same way you figured out what they like to eat, you can figure out where they like to live. Again, start by doing your research. Pick a species or two that live in your area and read about where they like to make their homes. For example, odorous house ants like to live in mulch, and winnow ants like to live in rotting logs.

Next, carefully inspect those areas. On ant hunts, I carry a small but sturdy garden spade to help with delicate digging and exploring. With winnow ants, for example, I like to go into the woods and carefully peel back the bark on rotting logs. It doesn't take long before I find a winnow ant colony in there, running about with their brood in their mandibles, trying to get away from me. I find many other species that way, like acrobat ants, carpenter ants, field ants, and citronella ants. In mulch around the bases of trees or next to houses, I often find Argentine ants or odorous house ants, as well as the occasional tiny thief ant or field ant colony. Sometimes I collect a bunch of acorns and crack them open to see who's living in there. In addition to acorn ants, I'll sometimes find Asian needle ants or odorous house ants. If you have ants living in your home, try to hunt them down by following their trails until you find a nest.

Be careful while you hunt for ants. Know which ants sting and always keep an eye out for other animals that could hurt you—like snakes, centipedes, or even raccoons— that might not want you poking around their homes. More than once, I've plunged my spade in a hollow log without looking only to be greeted by a cohort of angry yellow jackets, ready to tell me they don't like what I've done to their house. If I had paid closer attention, I would have noticed the wasps coming and going from their home and avoided a painful situation for me and my ant-hunting friends. Not paying attention can make you quite unpopular with your companions.

Keeping Ants as Pets

If watching ants in their natural environment isn't enough, and you still want to try your hand at keeping them at home, you have a number of ant-keeping options at your disposal. To get a better idea of what sort of housing might be appropriate, think about the type of ant you'd like to have and its preferred lifestyle. Ant houses work well for species that thrive in enclosed environments, but not others that like to forage over larger areas or up and down tree trunks.

Selecting Your Ants

As you know from reading this book, the world around you is bursting with ant species. Each has its own unique way of life, including diet and housing. You can purchase ants online from one of many ant retailers, but be warned that most retailers send several workers and no queen. Without the queen, the colony will not be able to replenish workers when they die, and the colony will die out in a few months to a year or so. Still, purchasing workers online can be an easy ant-acquiring solution and a rewarding experience for beginner ant keepers.

If you choose to capture your own ants, consider:

x   which ant species live around you and are active when you want to collect them
x   where those species live (in logs, underground, in mulch, under rocks, high in trees)
x   what those species eat (some diets may be more difficult to reproduce at home than others)
x   if those species are dangerous and put you or your household at risk. (You will probably want to avoid collecting Asian needle ants or red imported fire ants, for example, as they sting.)
x   if those species could become household pests. (When I studied ants in a laboratory, I accidentally infested my laboratory with Argentine ants more than once. Others do not appreciate this.)
x   if the species you select are invasive. Releasing invasive species into the environment, even accidentally, is illegal and can be harmful to the environment

Now that you've decided who you want to bring home, it's time to build or purchase your species' dream house. We'll collect the ants later, when we're ready.

Home, Sweet Home

Ant houses, called formicaria, can be purchased from a number of online retailers. Formicaria constructed from natural materials and customized for a particular ant species work best to ensure you'll be able to keep your ants for a long time. For those just getting started, I'm a fan of purchasing a good formicarium over building one at home. That way, somebody will mail to your house everything your ants need to get rolling, and you can see what works for others before you try it yourself.

When building an ant house, consider how the ant lives outdoors.

If you do decide to build your own ant house, remember that many species can escape containers easily, so you need to have containers with tight-fitting lids. These lids should be fitted with fine screens or have tiny ant-proof holes for proper ventilation.

Formicaria can take many different, beautiful forms.

Ants also need proper moisture. The simplest ant nest can be built using a test tube, a stopper with a hole in it large enough for the ants to crawl out, cotton, tin foil, and water.

Fill the test tube about three-quarters of the way with water. Stuff cotton in the tube with clean fingers until you feel the cotton moisten to your fingertip. You don't want the tube to leak, but you do want it to be moist. Then, put the stopper in the tube. Be sure there's enough room in the tube for your ants to crawl around. Wrap the tube in tin foil so the ants can have a dark environment. You can peel the foil back to observe your colony from time to time. This tube should be placed in your container with tight-fitting lid.

You can feed your ants by placing their food directly in the container. Check your test tube every week or so to ensure it stays moist. If it dries, add another test-tube nest. The ants will move in when they run out of water, and then you can remove their old nest.

Remember, ants do not like disturbances, so with a homemade ant house like this, you should only check on them once or twice a day at most. When you check on them, try to avoid breathing on them, as the carbon dioxide in your breath alarms them.

After you master this basic ant setup, you may want to modify your homemade formicarium to impress your friends, suit your ant-watching desires, or help your ants to live in a more "natural" environment.

If you want to build a home where you can watch your ants crawl around all day, check the Internet for some sample formicaria plans, which will provide you with step-by-step instructions on how to build your ant's dream home. You can also play around with expanding upon your existing setup.

Go Get Your Ants!

Although I do recommend purchasing a premade formicarium, I also recommend collecting the ants for that formicarium yourself. When you collect your own ants, you can get as many as you want, and you can ensure you have a mated queen that could potentially provide you with years of ant-watching pleasure.

The simplest way to collect ants is to go on an ant hunt, outlined above, armed with your spade and containers with tight-fitting lids. When you find the species you want, try to locate the queen, some workers, and a bit of brood. Queens are usually larger than the other ants in the nests and can often (but not always) be found near the brood. Delicately scoop up the queen, workers, and brood with a spade and place them in your container with a tight-fitting lid.

When you get home, place your test-tube nest in that container and wait for the queen and workers to move in. Leave them alone for a day or two so they feel comfortable.

Supper Time!

Now that you have a small colony, it's time to give them something to eat. Most ant species need protein, sugar, fats, and water. Read up on your species to see what types of food they like most. As you work to make fatties, keep in mind that just because they prefer sugary foods doesn't mean they don't need a little protein and fat every now and then. You can purchase ant food online, or you can give it a go yourself. Here is a list of foods I use:

PROTEIN
- ✕ Dead arthropods (sized appropriately for your species)
- ✕ Peanut butter (works sometimes but not all times)
- ✕ Seeds and nuts (for ants interested in seeds and nuts)
- ✕ Tuna fish in oil (they love it, but it smells bad)
- ✕ Pecan cookies (makes a nice snack but will not sustain the colony)

SUGAR
- ✕ Sugar water (50 percent sugar, 50 percent water)
- ✕ Honey mixed with a little water
- ✕ Apple jelly or some other fruit jelly mixed with a little water

FAT
- ✕ Tuna in oil has fats, as do seeds and peanut butter (above)
- ✕ Some people mix a little olive or canola oil with their sugar solutions

With all of these foods, it's important to remember ants' tiny stomachs. Because you want to avoid getting mold in your ant houses, try not to feed them too much, and remove uneaten food as it dries out or begins to mold. I like to feed them on tiny trays I fashion out of tin foil or in shallow bottle caps, so I can remove uneaten food easily.

Watching Your Ants

Now that you have ants in your living room, spend some time watching them to see what they do. How do they eat? When you put out new food for them, how do they find it? How long does it take them to find it? Which foods do they pick up most often and most readily? Which foods do they avoid? What do they do in their nest chambers all day? Does light bother them? What times of year does the queen lay eggs? How do they feed their larvae? Watching ants can be fascinating. You'll be amazed how quickly the time goes.

PRO TIPS FROM A PROFESSIONAL ANT COLLECTOR

Keeping ants at home can be complicated. Mack Pridgen, founder of the ant house business Tar Heel Ants, has figured out some valuable tricks to the trade, and he wants to share them with you.

DR. ELEANOR: How did you get interested in keeping ants?

MACK: I was fascinated by ants as a kid. We had some very large colonies of carpenter ants (*Camponotus*) around my house, and I was the kid with the jars trying to collect them and feed them. My favorites were the giant red ants with big heads! I now know those as the carpenter ant *Camponotus castaneus* majors.

Years later I bought my daughter a gel ant farm as a Christmas gift and decided to show her how cool ants were to watch digging. Instead of ordering ants (when I was a kid, mine arrived dead), I went out the day after Christmas and found carpenter ant *Camponotus chromaiodes* workers foraging up a tree. They died within a couple of days. I pressed on, researching how people cultivate ants in labs, and I reached out to the local ant lab here in town. Before long I was raising young colonies in test tubes and planning their first formicarium!

DR. ELEANOR: What common ants would you say are the "best" (easiest, safest) for beginners to keep at home?

MACK: It depends on where you live. Concentrated in the southeastern United States but also found in other regions throughout the country, various winnow ants are abundant in forests under rocks. Winnow ants are always my first suggestion. I was lucky enough to have them for my first large colony.

Carpenter ants (*Camponotus* spp.) also work well for beginners. Their queens are huge, around three-quarters of an inch in size, and they vary in color.

DR. ELEANOR: What do you find to be some of the common mistakes people make when keeping ants at home?

MACK: Common mistakes include

1. **Improper feeding expectations and procedures.** For example, most queens do not need food until their first workers have emerged. The workers find food and feed the queen. Many inexperienced ant keepers try to feed ants live insects, too much of a dead insect, or sweet liquids inside their claustral chambers (test tubes or other starting formicaria). This stresses the queen, causes mold problems for the queen and her brood, and more. Carefully feeding small amounts of foods is OK but the excess should be removed immediately. Practicing a proper feeding regimen can make all the difference for your colony, and an improper regimen is often an ant keeper's biggest problem. Follow these proper feeding techniques:

   × Offer small pieces of fresh insects often—at least a couple times a week—and anchor them to something (cork, silicon, etc.) to prevent trash from entering the formicarium (less important if you can disassemble your formicarium).

   × Always provide liquids and solids on dishes, regardless of colony size. Any food that smears on the formicarium surfaces or foraging area can mold and lead to major problems for the colony down the road. Providing some loose (ant-safe) substrate allows your ants to cover the problem areas themselves.

× Never feed your colonies live food. Workers can get injured battling living insects. At minimum, injure prey if live feeding is a must for your particular species.

× Rotate foods. Use a combination of fresh fruits (apples are a personal favorite), dead crickets, mealworms, and fruit flies, in addition to other foods that provide a good balance of carbohydrates, proteins, and fats.

2. **Not anticipating how easily ants can climb and escape.** One of the first major considerations while planning your homemade formicarium is security. Think of this as protection for your ants, not you!

3. **Not planning ahead.** Colonies can grow fast during their growth season. Many of the species we see in urban areas, such as pavement ants, fire ants, and acrobat ants, can quickly outgrow their formicaria and can be difficult to contain and feed if not kept in a proper-size habitat.

DR. ELEANOR: What are your top tips for building an ant house at home?
MACK:

1. **Use what you already know.** You can build an ant home using skills you already have. Good with crafts? Perhaps build a plastic nest out of acrylic or make a plaster nest. Artistically inclined and like to carve? Good with power tools? Try finding some Ytong (AAC block) bricks and crafting your own design.

2. **Plan your hydration first.** Moisture is your colony's lifeline. If you can, use a hygrometer to test the moisture level in your formicarium before introducing your ants.

3. **Ask questions.** Many people have been where you are now. Join an online group or forum.

4. **Keep local ants!** Don't risk spreading harmful species.

DR. ELEANOR: What are your top five tips for keeping ants alive at home?

MACK:

1. **Monitor the temperature** near your colonies daily with a thermometer. Keep your ants at a constant temperature (78–82°F, typically, though some species may prefer warmer conditions). Quick temperature changes can be problematic for ants. Remember: Ants are not plants. They don't do well in the sun!
2. **Label your ant colonies** in case friends or families don't know what they are. Shaking or tilting your formicarium is not recommended.
3. **Wear disposable latex gloves** or, at a minimum, wash your hands before feeding your ants or handling their formicaria.
4. **Use organic fruits and honey** when possible.
5. **Clean out your ant foraging areas regularly**, whenever you see detritus build up.

ACKNOWLEDGMENTS

This book would not have been possible without the expertise and help of many great individuals. Thank you, Rob Dunn and Andrea Lucky, for having the curiosity and vision to help us all turn over stones and peer in the grass to find the ants around us. Rob, thank you, also, for your editorial guidance, opportunities, and encouragement. You and the University of Chicago Press have opened the door for us to share our joy of ants. My great appreciation goes to the University of Chicago Press's Christie Henry, Gina Wadas, Logan Ryan Smith, and Amy Krynak for their guidance and support. Robin Anders, you know how to edit and critique like nobody else. Thank you. Thank you also, Kathryn and Jamesie Spicer, for your editorial assistance and for giving me the crumbs and long mornings and afternoons to meet my ants. Neil McCoy, thank you for your creativity, which helps to sharpen and enliven ideas. Holly Menninger, your powers of coordination and organization are unparalleled. Thank you, Alex Wild, for using your skill and vision to make giants of ants, for showing us how beautiful their tiny world is and how they can be our friends. Thank you, E. O. Wilson and Corrie Moreau, for your words and your discoveries. Thank you for your expert advice and encouragement, Matt Shipman. Russ Campbell and the Burroughs Wellcome Fund, thank you for recognizing this book and giving it a chance. Thank you to my ant people, including Jules Silverman, Alexei Rowles, John Brightwell, Bill Reynolds, Sean Menke, Jon Shik, Brad Powell, David Bednar, Grzesiek Buczkowski, Heike Meissner, Benoit Guenard, Clint Penick, and Amy Savage, plus one snake guy, Warren Booth. Greg Rice, you are the words and the ants and everything else. Thank you.

GLOSSARY

abdomen: The third major division of the insect body (aka rump, booty, posterior, etc.) that contains most of the ant's organs and its stinger.

ant: A small, wingless, wasp-like insect that usually lives in eusocial groups. An incredibly diverse and ecologically important animal. Earth houses more than 15,000 named ant species, and many more awaiting scientific description and naming. *See* Hymenoptera and Formicidae.

antenna (pl. antennae): A segmented appendage projecting from either side of an adult insect's head. Antennae function as sensory organs and help ants sniff, feel, and taste.

aphid: A small, plant fluid–sucking insect that usually resembles a tiny cicada or a tiny, chubby katydid. Aphids can be winged or wingless and usually are found on the undersides of plant leaves or along stems. Often protected by many ant species, aphids turn excess plant fluid into a sweet substance called honeydew, which ants eat.

arthropod: An animal with jointed legs and an exoskeleton. Arthropoda refers to the large scientific group including shellfish, insects, scorpions, and spiders. *Arthro* comes from the Greek word meaning "joint," and *poda* comes from the Greek word meaning "foot." The vast majority of described species on earth are arthropods. Ants are arthropods.

biodiversity: The amount of different life forms in an area. In general, a rich biodiversity (lots of different life forms) means a healthy environment. Some invasive ant species, like Asian needle ants, reduce biodiversity when they move into an area, which could result in an unhealthy habitat.

caste: Refers to the various groups of ants within a colony. Sexual castes consist of two groups: males and females. Morphological castes consist of two or more groups, typically minors and majors (soldiers). Temporal castes divide ants according to age and the jobs they do at those ages. Reproductive castes refer to queens, which reproduce, and workers, which don't.

colony: A group of ants, often closely genetically related, which operate as a functional unit without aggression among the group. Colonies can have one or many nests and one or many queens.

common name: The moniker we call ants for convenience. Most people referring to ants use their common names. Common names usually refer to some aspect of the ant's appearance (like "little black ant") or behavior (like "crazy ant"). Common names can be different in different languages.

complete metamorphosis: A form of insect development, in which the insect undergoes the following stages to adulthood: egg larva (looks very different from adults) pupa adult. Ants undergo complete metamorphosis.

crop: A "stomach" attached to the esophagus that serves to receive and hold food. It's like an internal backpack. Crops hold food without digesting it so ants can share it with their sisters or eat it later.

ecology: The study of the relationships between living things and their environment. *Eco* comes from the Greek word for "house," and *ology* comes from the Greek word meaning "the study of."

egg: The first stage in an ant's development and laid by queens, an egg has a simple germ cell, nutritious yolk, and a surrounding membrane.

entomology: The study of insects and other arthropods. *Entom* comes from the Greek word for "insect," and *ology* comes from the Greek word meaning "the science of."

entomologist: Someone who studies insects and other arthropods.

eusocial: If an animal cooperatively cares for its young, has a reproductive division of labor (for example, queens reproduce; workers work), and an overlap of at least two generations sharing a space and contributing to the group, then it's eusocial. Most ant species are eusocial. The few that are not eusocial are workerless parasites in the nests of other ants.

exoskeleton: The "hard outer shell" of insects and other arthropods. Instead of bones, insects have a suit of armor consisting of a plastic-like substance called chitin, which is covered by a thin layer of waxy material. Ant muscles are attached on the inside of the exoskeleton.

exotic: In invasion ecology, exotic refers to an organism that is present in an area but which comes from a different place. That is, that organism did not evolve in that area. Exotic species are not always invasive, and they're not always pests. Your pet cat is an exotic species, and some would say it is a pest. Honey bees are also exotic species in the United States. They come from Europe and Africa.

Formicidae: The scientific grouping called family to which all ants belong. The word *Formicidae* comes from the Latin word meaning "ant."

gaster: The swollen part of the abdomen behind the ant's skinny waist, or petiole.

genus (pl. genera): A group of species that share characteristics and are often closely related. For example, thief ants and red imported fire ants share many physical characteristics and are closely related. They share the genus *Solenopsis.* Knowing genera can help you mentally group ants by form and function.

holometabolous: The quality of an organism, like an ant, of undergoing complete metamorphosis.

honeydew: A sugary fluid excreted from the abdomens of many different insects, including aphids and scale insects. Many ant species love to eat honeydew and rely on it for survival.

Hymenoptera: The scientific order of insects to which ants belong. Bees and wasps also belong to this order, and these three types of insects share much in common, including their skinny waists and their tendency toward forming social groups. Hymenoptera comes from *hymen*, the Greek word for "membrane," and *ptera*, the Greek word for "wings." Hymenopteran wings look like a thin membrane stretched across a few veins.

insect: A class of animal that has an exoskeleton, three major body segments (head, thorax, and abdomen), six legs, and two antennae. Ants are insects. Spiders (eight legs, two segments) are not.

invasive species: A species that moves into an area and negatively impacts that environment. Red imported fire ants and Asian needle ants are examples of invasive species.

invertebrate: A general term referring to any animal that does not have a backbone. Worms, insects, crabs, octopi, and spiders are all invertebrates. Most of life on earth has no backbone.

larva (pl. larvae): The second stage in an ant's development, between egg and pupa. Larvae differ in form from adults. Ant larvae often look like legless grubs.

major: A worker subcaste (*see morph*) in which the individual is typically larger and specialized for defense. Big headed ants have the most prominent majors, but other ant species, like carpenter ants, can have majors, too. Sometimes referred to as soldiers.

mandible: The first pair of jaws in ants. Mandibles usually stick out from the front of the head and are good for chomping, slicing, and carrying.

minor: A worker subcaste (*see morph*) in which the individual is typically smaller and specialized for work.

morph: Any of the various forms of ants within a caste. For example, a major is one morph, while a minor is another.

myrmecologist: A person who studies ants.

native species: An organism that is present in an environment "naturally" and not because a human facilitated its presence in the environment.

nest: Among ant species, a nest is a discrete living space for a related group, usually containing workers, brood, and queens, but sometimes containing any two of the three (or any combination plus males). Nests can be as simple as a hangout spot under a rock (as with odorous house ants) or as complex as intricate underground tunnel networks connecting rooms (as with winter ants). Ant species can have one or many nests per colony.

nestmate: Individuals, usually related, who share a nest. Nestmate can also refer to members of the same multinest colony who don't share a particular nest. Nestmates do not fight one another when they meet outside the nest, and they recognize one another as nestmates because they smell alike.

pest: A species that negatively impacts its environment. Some ant pests, like odorous house ants, are "nuisance pests," meaning that people tend to be bothered by them but they don't necessarily negatively impact the environment. The primary damage associated with these pests is indirect, resulting from people's use of hazardous chemicals to exterminate them. Other pest ants, like Asian needle ants, are true pests, meaning they cause economic damage (like crop loss), or pose health risks (as from the sting of an Asian needle ant). Not all pests are invasive species or exotic species, and not all exotic species are pests.

petiole: The skinny segments at the beginning of the abdomen, between the thorax and gaster, that give ants their skinny, wasp-like "waists."

pheromone: Any one of many chemical secretions used to communicate within species. Ants use a variety of pheromones to communicate, including alarm pheromones, recognition pheromones, and trail pheromones.

polyethism: The division of labor among members in the colony. Different forms of polyethism are apparent in ant colonies. For example, many ants display something called age-based polyethism, where younger workers perform different tasks than older workers.

polymorphism: In ants, having several physical forms of adults. Many insects display polymorphism.

pupa (pl. pupae): The life cycle stage in insects with complete meta-morphosis. In ants, it occurs between the larva and adult stages, when the insect becomes inactive, doesn't eat, and develops the physical features of an adult.

queen: In ants, female colony members who can lay fertilized eggs. Usually larger than workers.

scale insect: A small, plant fluid–feeding insect that looks like a bump, shell, or scale stuck to plant bark or stems. Often protected by many ant species, scale insects turn excess plant fluid into a sweet substance called honeydew, which ants eat.

scientific name: The formal epithet used to describe species; regulated by a huge international formal naming process. Usually with Greek or Latin roots, scientific names are the same in all languages across the globe. This standardization is extremely useful for communicating science. Just like we have first and last names, scientific names consist of two parts: one for genus and the other for species. As knowing somebody's name can tell you about that person, knowing scientific names can tell you a lot about the insect. The genus name is like our last name and the species name is like our first name. For example, my name is Eleanor Spicer Rice. If I told somebody from my hometown my name, that person would know I'm kin to the Spicers and could have a general knowledge about me before they even got to know me. If she knew my relatives, she could get an idea of what I might look like and could have an idea of where I live and to a certain degree how I might behave. If you tell an ant scientist you saw a *Brachyponera chinensis*, even if he's never met one, he would know a lot about how the species looks, lives, and acts if he knows other *Brachyponera*. Just as my first name, Eleanor, distinguishes me from the other Spicers hanging around town, the specific epithet distinguishes each species from all the other species and gives us an idea of what that species does. For our *Brachyponera chinensis*, "chinensis" tells us this species is native to Asia. While species might have the same specific name, no two species share both genus and species name. That way, there's no confusion about which species scientists are talking about.

segment: In insects, any division of the body. While segment can refer to each joint in the leg or antenna, we most often think of segments when discussing one of the three major insect body divisions: head, thorax, and abdomen.

soldier: See major.

species: A group of individuals that are genetically similar and able to mate and produce offspring that can also mate and produce offspring.

spiracle: The holes on an insect's body that open to its respiratory, or tracheal, system. Basically, it's how the insect breathes. Like our mouth or nose.

thorax: The second, or middle, segment of an insect. The thorax is the locomotion center.

trophallaxis: In ants and other eusocial insects, the process of exchanging crop contents between individuals through the mouth. It's one way ants share food and communicate information.

worker: In social insects like ants, a member of the laboring caste that isn't able to reproduce.

ADDITIONAL RESOURCES

Throughout the book, we relied on a collection of excellent resources that we hope will help you, too, as you continue your love affair with ants.

WEBSITES

× Antweb (www.antweb.org) offers the world's largest ant-centric database, complete with photographs, cool ant stats, and the latest research from curators across the country.

× Alex Wild snaps up-close photos of thousands of ants and displays them on www.alexanderwild.com.

× Joe MacGown's beautifully illustrated ant keys may help you identify your formicid neighbors. Explore his findings, plus a bunch of other great insect information, on the Mississippi Entomological Museum website (http:// mississippientomologicalmuseum.org.msstate.edu).

× With AntWiki (www.antwiki.org) and AntCat (www.antcat.org), you can check out all the ant taxonomic information you could ever want.

× And if you ever want to read scientific literature about ants, a great place to start is the USDA's FORMIS page (http://www.ars.usda.gov/News/docs. htm?docid=10003).

BOOKS

× *Ants of North America: A Guide to the Genera*, by Brian L. Fisher and Stephan P. Cover, is the best ant field guide around. In addition to the helpful and beautiful ant graphics, the authors give you an engaging natural history of each genus, helping you to get to know these ants on a deeper level.

× *Journey to the Ants. A Story of Scientific Exploration*, by Bert Hölldobler and E. O. Wilson is an exciting, beautiful book about the discovery and love of our natural world (through the eyes of two of the world's best and most treasured ant lovers).

× *Adventures among Ants. A Global Safari with a Cast of Trillions*, by Mark Moffett. The "Indiana Jones of Ants" shares personal stories and interesting ant behaviors around the world in this page turner.

× *The Ants*, by Bert Hölldobler and Edward O. Wilson, is the definitive guide to all things ants. Despite its intimidating size, this book is an engaging read and illuminates every ant nook and cranny, from ant evolution to their array of behaviors to their intricate physiologies and taxonomic details.